Nehemiah Grew

The Anatomy of Vegetables begun

With a general Account of Vegetation founded thereon

Nehemiah Grew

The Anatomy of Vegetables begun
With a general Account of Vegetation founded thereon

ISBN/EAN: 9783337163297

Printed in Europe, USA, Canada, Australia, Japan

Cover: Foto ©berggeist007 / pixelio.de

More available books at **www.hansebooks.com**

THE ANATOMY OF VEGETABLES Begun.

With a

GENERAL ACCOUNT OF *VEGETATION* Founded thereon.

By *NEHEMIAH GREW*, M. D.
and Fellow of the *Royal Society*.

LONDON,
Printed for *Spencer Hickman*, Printer to the *R. Society*, at the *Rose*
in S. *Pauls* Church-Yard, 1672.

TO THE

Right Honourable

&

Most Illustrious

THE

PRESIDENT & FELLOWS

OF THE

ROYAL SOCIETY,

The Following

DISCOURSE

Is most Humbly

Presented

By

The Authour

NEHEMIAH GREW.

TO THE

Right Reverend

JOHN

Lord Bishop of

CHESTER.

MY LORD,

I hope your pardon, if while you are holding *that Best of Books* in one Hand, I here present some Pages of

 that

that of *Nature* into your other: Eſpecially ſince *your Lordſhip* knoweth very well, how excellent a *Commentary* This is on the *Former*; by which, in part God reads the World his own Definition, and their Duty to him.

But if this Addreſs, *my Lord*, may be thought congruous, 'tis yet more juſt; and that I ſhould let *your Lordſhip*, and others know, how much, and how deſervedly I reſent your extraordinary Favours: Particularly that you were pleaſed ſo far to animate my Endeavours towards the publiſhing the following *Obſervations.* Many whereof, and moſt belonging to the Firſt Chapter, having now lain dormant near ſeven years; and yet might perhaps have ſo continued, had not *your Lordſhips* Eye at length created

created Light upon them. In doing which, you have given one, amongſt thoſe many Tokens, of as well your readineſs to promote learning and knowledge by the hands of others; as your high Abilities to do it by your own: Both which are ſo manifeſt in *your Lordſhip*, that like the firſt Principles of *Mathematical Science*, they are not ſo much to be aſſerted, because known

and

and granted by all.

The Consideration whereof, *my Lord*, may make me not only *just* in owning of your Favours, but also most *Ambitious* of your *Patronage*: which yet to bespeak, I must confess I cannot well. Not that I think what is good and valuable, is alwaies its own best Advocate; for I know that the Censures of men are humorous and variable, and that one

Age

Age must have leave to frown on those Books, which another will do nothing less than kiss and embrace. But chiefly for this Reason, lest I should so much as seem desirous of *your Lordships* Solliciting my Cause as to all I have said: For as it is your Glory, that you like not so to shine, as to put out the least Star; so were it to your Dishonour to borrow your

Name to illuſtrate the Spots, though of the moſt conſpicuous.

Your Lordſhips

Moſt Obliged

&

Moſt Humble

Servant

Nehemiah Grew.

THE PREFACE.

OF *what antiquity the* Anatomy *of* Animals *is, and how great have been its Improvements of later years, is well known. That of* Vegetables *is a Subject which from all Ages to this day hath not only lain by uncultivated; but for ought I know, except ſome Obſervations of ſome of our own Countrey-men, hath not been ſo much*

as

as thought upon; whether for that the World hath been more enamoured with the former, or pity to humane frailty hath more obliged to it, or other Reasons, I need not enquire.

But considering that both came at first out of the same Hand, and are therefore the Contrivances of the same Wisdom; I thence fully assured my self, that it could not be a vain Design, though possibly unsuccessful, to seek it in both.

In the prosecution hereof, how far I have gone, I neither judge my self, nor leave it to any one else to do it; because no man knows how far we have yet to go, or are capable of going. Nor is there any thing which starves and stinteth the growth of knowledge more, than such Determinations, whether we speak or conceit them only.

What we have performed thus far, lieth, for the most part, open to the

use

use and improvement of all men. Only in some places, and chiefly in the Third Chapter, we have taken in the help of Glasses; wherein, after we had finished the whole Composure, some Observations made by that Ingenious and Learned Person Mr. Hook, *a Worthy Member of the* Royal Society, *my much Honoured Friend, and by him communicated to me, were super-added: As likewise some others also* Microscopical, *of my own, which his gave me the occasion of making.*

Those that shall think fit to examine, as well as to peruse these Observations, we advertise them, First, *That they begin, and so proceed till they end again, with the Seed: For they will hardly be able to avoid Errour and Misapprehension, if either partial or preposterous in their Enquiries.* Next, *That they confine not their Enqniries to one time of the Year; but to make them in several Seasons, wherein the Parts of a* Vegetable *may be seen in their several Estates.* And then, *That they neglect not the comparative* Anatomy; *for as some things are better seen in one estate, so in one* Vegetable, *than another.*

What, upon Observation already made, we have erected, as they are not Sticks and Straws; so neither do we assure all to be of the best Oak. How Dogmatical soever my Assertions may seem to be, yet do I not affect the unreasonable Tyranny of obtruding upon the Faith of any. He that speaketh Reason, may be rather satisfied, in being understood, than believed.

THE

THE

CONTENTS

CHAP. I.

Of the Seed as Vegetating.

CHAP.

CHAP. 2.

Of the Root.

CHAP. 3.

Of the Trunk.

The Appendix.

Of Trunk-Roots and Claspers.

CHAP.

CHAP. 4.

Of the Germen, Branch, and Leaf.

The Appendix.

Of Thorns, Hairs and Globulets.

CHAP. 5.

Of the Flower.

CHAP. 6.

Of the Fruit.

CHAP. 7.

Of the Seed in its ſtate of Generation.

mulets

Cl. Glissonius in Prolegomenis præfixis Libro de *Hepatis Anatomia*, c. 1.

Plantæ quoque in hunc censum (*sc. Anatomicum*) veniunt. Varia enim partium textura, & differentiis constant: & proculdubio ex accurata earundem dissectione, utiles valde Observationes nobis exurgerent; præstaretq; in illis (inferioris licet ordinis) rebus examinandis operam impendere, re, quam in tranſcribendis (ut ſæpe fit) aliorum laboribus, inutiliter ætatem tranſigere. Quippe, hoc pacto, ignavarum apum more, aliena duntaxat alvearia expilamus, nihilq; bono publico adjicimus.

To be added and corrected.

PAg. 8. *l.* 15. after *must*, *adde* upon the Sprouting of the *Bean*. *p.* 12. *l.* 23. after *dense*, *adde* and thence their different Tinctures. *p.* 18. *l*, 13. after *that*, *adde* when. *p.* 20. *l.* 8. for *the*, *read* an. *p.* 56, *l.* 8. *r.* once. *p.* 90. *l.* 11. *dele* as. *p.* 91. *l.* 12. *r.* older. *p.* 120. *l.* 11. after *all*, *r.* is. *p.* 134. *l.* 11. *r.* *Convolvulus*. *p.* 143. *l.* 10. *r.* ever. *p.* 145. *l.* 14. for *not*, *r.* or. *p.* 159. *l.* 8. for *by*, *r.* to. *p.* 160. *l.* 18. dele *not*. *p.* 185. *l.* 14. after *therewith*, *r.* the. *dele* the former *the*.

In some Copies.

P. 168. *l.* 4. *r.* *ultimate end*, and *p.* 170. *l.* 22. *r.* *Favours*.

The Reader is desired to excuse the misplacing of the Figures by the Graver, in the Authors absence.

THE ANATOMY OF VEGETABLES Begun.

With a General Account of *Vegetation* founded thereon.

CHAP. I.

Of the Seed as Vegetating.

BEing to ſpeak of Vegetables; and, as far as Inſpection and conſequent Reaſon may conduct, to enquire into the viſible Conſtitutions and Uſes of their ſeveral

Parts; I chuſe that Method which may with beſt advantage ſuit to what we have to ſay hereon: And that is the Method of Nature her ſelf, in her continued Series of Vegetations, proceeding from the Seed ſown, to the formation of the Root, Trunk, Branch, Leaf, Flower, Fruit, and laſt of all, of the Seed alſo to be ſown again; all which we ſhall in the ſame order particularly ſpeak of.

The Eſſential Conſtitutions of the ſaid Parts are in all Vegetables the ſame: But for Obſervation, ſome are more convenient; in which I ſhall chiefly inſtance. And firſt of all, for the Seed we chuſe the great Garden-Bean.

If we take a Bean then and diſſect it, we ſhall find it cloathed with a double Veſt or Coat: Theſe Coats, while the Bean is yet green, are ſeparable, and eaſily diſtinguiſhed. When 'tis dry, they

they cleave ſo cloſely together, that the Eye, not before inſtructed, will judge them but one; the inner Coat likewiſe (which is of the moſt rare contexture) ſo far ſhrinking up, as to ſeem only the roughneſs of the outer, ſomewhat reſembling Wafers under *Maquaroons*.

At the thicker end of the Bean, in the outer Coat, a very ſmall *Foramen* preſents it ſelf: In diſſection 'tis found to terminate againſt the point of that part which I call the *Radicle*, whereof I ſhall preſently ſpeak. It is of that capacity as to admit a ſmall Virginal Wyer, and is moſt conſpicuous in a green Bean.

This *Foramen* may be obſerved not only in the great Garden-Bean, but likewiſe in the other kinds; in the French-Bean very plainly; in Peaſe, Lupines, Vetches, Lentiles, and other Pulſe 'tis alſo found: and in many Seeds not reckoned of this kindred, as in that of *Fœnugreek*, *Medica Tornata*, *Goats-Rue*, and others: In many of which, 'tis ſo very ſmall, as ſcarcely, without the help of Glaſſes to be diſcovered; and in ſome, not without cutting off part of the Seed beſides, which otherwiſe would intercept the ſight hereof; it being in theſe and ſuch like Seeds, from the place of the breaking off of the Peduncle perfectly diſtinct.

We may then obſerve, that all Seeds which have thick or hard Coats, have the ſame likewiſe perforated, in this, or ſome other manner. And accordingly, although the Coats of ſuch Seeds as are lodg'd in Shells or Stones, being thin, are not viſibly perforated; yet the Stones and Shells themſelves always are; as *Chap.* 7. ſhall be ſeen how. To which Chapter, what is farther obſervable, either

as

as to the nature, or number of the covers of the Seed, I alſo refer.

The Coats of the Bean being ſtripp'd off, the proper Seed ſhews it ſelf. The parts whereof it is conſtituted, are three; *ſc.* the main Body, and two other appendant to it, which we may call the three Organical parts of the Bean.

The main Body is not one entire piece, but alwaies divided lengthwiſe into two halves or Lobes, which are both joyn'd together at the Baſis of the Bean. Theſe Lobes in dry Beans, are but difficultly ſeparated or obſerv'd; but in young ones, eſpecially boil'd, they eaſily ſlip aſunder. See *Fig.* 1.

Some very few Seeds are divided, not into two Lobes, but more; as that of *Creſſes*; and ſome not at all divided, but entire; as *Corn*: Excepting which few, all other Seeds, even the ſmalleſt are divided, like as the Bean, into juſt two Lobes: whereof, though in moſt Seeds we cannot by diſſection be inform'd; yet otherwiſe we eaſily may as ſhall be ſeen.

At the Baſis of the Bean, the two other Organical parts ſtand appendent; by mediation whereof the two Lobes meet and join together. The greater of theſe two parts ſtands without the two Lobes, and upon diveſting the Bean of its Coats, is immediately viſible. 'Tis of a whiter colour, and more gloſſie than the main Body, eſpecially when the Bean is young. In the Bean, and many other Seeds, 'tis ſituated ſomewhat above the thicker end, as you hold the Bean in its moſt proper poſture for growth. In Oak-Kernels, which we call Acorns, Apple-Kernels, Almonds, and many other Seeds, it ſtands prominent juſt from the end; the Baſis and the end being in in theſe the ſame, but in the Bean divers. See *Fig.* 1.

This part is not only in the Bean, and the Seeds above mentioned; but in all others: being that which upon the Vegetation of the Seed, becomes the Root of the Plant; which therefore I call the *Radicle*: by which, I mean the Materials, abating the Formality, of a Root. 'Tis not eaſie to be obſerved, ſaving in ſome few Seeds, amongſt which, that of the Bean is the moſt fair and ample of all I have ſeen; but that of ſome other Seeds, is, in proportion, greater; as of *Fœnugreek*, which is almoſt as big as one of its Lobes.

The leſſer of the two ſaid Appendents lies occult between the two Lobes of the Bean, by ſeparation whereof only it is to be ſeen. 'Tis enclos'd in two ſmall Cavities form'd in the Lobes for its reception. Its colour comes near that of the *Radicle*; and is founded upon the Basis thereof, having a quite contrary production, *sc.* towards the cone of the Bean; and being that very part, which, in process, becomes the Body or Trunk of the Vegetable. See *Fig.* 1.

For the sake of this Part principally it is, that the Bean is divided into Lobes; *sc.* that it may be warmly and safely lodged up between them; and so secur'd from the Injuries so tender a Part would sustain from the Mould, whereto, had the Main Body been entire, it must have lain contiguous.

This Part is not, like the *Radicle*, an entire Body, but divided at its loose end into divers pieces, all very close set together, as Feathers in a Bunch; for which reason it may be called the *Plume*. They are so close, that only two or three of the outmost are at first seen: but upon a nice and curious separation

ſeparation of theſe, the more interiour ſtill may be diſcovered. Now as the *Plume* is that Part which becomes the Trunk of the Plant, ſo theſe pieces are ſo many true, and already formed, though not diſplayed, Leaves, intended for the ſaid Trunk, and foulded up in the ſame plicature, wherein, upon the ſprouting of the Bean, they afterwards appear. In a French Bean the two outmoſt are very fair and elegant. In the great Garden-Bean, two extraordinary ſmall Plumes, often, if not always, ſtand one on either ſide the great one now deſcrib'd: From which, in that they differ in nothing ſave in their ſize, I therefore only here juſt take notice of them. And theſe three Parts, *ſc.* the *Main Body*, the *Radicle*, and the *Plume*, are concurrent to the making up of every Seed; and no more than theſe.

Having thus taken a view of the Orga-

Organical Parts of the Bean, let us next examine the Similary, *sc.* those whereof the Organical are compos'd: a distinct observation of which, for a clear understanding of the Vegetation of the Seed, and of the whole Plant arising thence, is requisite: To obtain which, we must proceed in our Anatomy.

Dissecting a Bean then, the first Part occurring is its Cuticle. The Eye and first Thoughts suggest it to be only a more dense and glossy Superficies; but better enquiry discovers it a real Cuticle. 'Tis so exquisitely thin, and for the most part so firmly continuous with the Body of the Bean, that it cannot, except in some small Rag, be distinctly seen; which, by carrying your Knife superficially into the Bean, and then very gently bearing upward what you have cut, will separate and shew it self transparent.

parent. This Cuticle is not only ſpread upon the Convex of the Lobes, but alſo on their Flats, where they are contiguous, extending it ſelf likewiſe upon both the *Radicle* and *Plume*, and ſo over the whole Bean.

This Part, though it be ſo far common with the Coats of the Bean, as to be like thoſe, an Integument; yet are we in a quite different Notion to conceive of it: For whereas the Coats upon ſetting the Bean, do only adminiſter the Sap, and, as being ſuperſeded from their Office, then die; as ſhall be ſeen: this, on the contrary, with the Organical Parts of the Bean, is nouriſhed, augmented, and by a real Vegetation co-extended.

Next to the Cuticle, we come to the *Parenchyma* it ſelf; the Part throughout which *the inner Body*, whereof we ſhall ſpeak anon, is diſſeminated; for which reaſon

I call it the *Parenchyma*. The Surface hereof is ſomewhat denſe, but inwardly 'tis more porous, and of a laxer Contexture. If you view it in a Microſcope, it hath ſome ſimilitude to the Pith, while ſappy, in the Roots and Trunks of Plants; and that for good reaſon, as in *Ch*. 2. ſhall be ſeen, This is beſt ſeen in green Beans. See *Fig*. 2.

This Part would ſeem by its colour to be peculiar to the Lobes of the Bean; but as is the Cuticle, ſo is this alſo, common both to the *Radicle* and *Plume*; that is, the *Parenchyma* of the Bean, as to its eſſential ſubſtance, is the ſame in all three. The reaſon why the colour of the *Plume*, and eſpecially of the *Radicle*, which is white, is ſo different from that of the Lobes, may chiefly depend upon their being more compact and denſe. And therefore the Lobes themſelves, which are green while the Bean is

young;

young; yet being old and dry, become whitiſh too. And in many other Seeds, as Acorns, Almonds, the Kernels of Apples, Plums, Nuts, &c. the Lobes, even freſh and young, are pure white as the Radicle it ſelf.

But although the *Parenchyma* be common, as is ſaid, to all the Organical Parts; yet in very differing proportions. In the *Plume*, where it is proportionably leaſt, it maketh about three Fifths of the whole *Plume*; in the *Radicle*, it maketh about five Seavenths of the whole *Radicle*; and in each Lobe, is ſo far over-proportionate, as to make at leaſt nine Tenths of the whole Lobe.

By what hath been ſaid, that the *Parenchyma* is not the only conſtituting Part, beſides the Cuticle, is imply'd: there being anothet Body, of an eſſentially different ſubſtance, emboſom'd herein: which

which may be found, not only in the *Radicle* and *Plume*, but alſo in the Lobes themſelves, and ſo in the whole Bean. See *Fig.* 2.

This inner Body appears moſt plain and conſpicuous in cutting the *Radicle* athwart, and ſo proceeding by degrees towards the *Plume*, through both which it runneth in a large and ſtraight Trunk. In the Lobes, being it is there in ſo very ſmall proportion, 'tis difficultly ſeen, eſpecially towards their Verges: yet if with a ſharp Knife you ſmoothly cut the Lobes of the Bean athwart, divers ſmall Specks, of a different colour from that of the *Parenchyma*, ſtanding therein all along in a Line, may be obſerv'd; which Specks are the Terminations of the Branches of this inner Body. See *Fig.* 3.

For this inner Body, as it is exiſtent in every Organical part of the Bean; ſo is it, with reſpect to each

part,

part, moſt regularly diſtributed. In a good part of the *Radicle* 'tis one entire Trunk; towards the Baſis thereof, 'tis divided into three main Branches; the middlemoſt runneth directly into the Plume; the other two on either ſide it, after a little ſpace, paſs into the Lobes; where the ſaid Branches dividing themſelves into other ſmaller; and thoſe into more, and ſmaller again, are terminated towards the Verges of each Lobe; in which manner the ſaid inner Body being diſtributed, it becomes in each Lobe, a true and perfect Root. See *Fig.* 2.

This Seminal Root, as now we'll call it, being ſo tender, cannot be perfectly excarnated, as may the Veſſels in the Parts of an Animal, by the moſt accurate Hand; yet by diſſection begun and continu'd, as is above-declared, its whole frame and diſtribution may be eaſi-

ly

ly obſerv'd. Again, if you take the Lobe of a Bean, and length-wiſe pare off its *Parenchyma* by degrees, and in very thin Shives, many Branches of the Seminal Root, (which by the other way of Diſſection were only noted by ſo many Specks) both as they are fewer about the Baſis of the Bean, and more numerous towards its Verges, in ſome good diſtinction and entireneſs will appear. For this you muſt have new Beans.

As the inner Body is branched out in the Lobes, ſo is it in the *Plume*: For if you cut the *Plume* athwart, and from the Baſis proceed along the Body thereof, you'l find therein, firſt, one large Trunk or Branch, and after four or five very ſmall Specks round about it, which are the terminations of ſo many leſſer Branches therewith diſtributed to the ſeveral parts of the *Plume*. See *Fig.* 4. The diſtribution

ſtribution of the inner Body, as it is continuous throughout all the Organical Parts of the Bean, is repreſented by *Fig.* 2.

This *Inner Body* is, by diſſecti-on, beſt obſervable in the Bean and great Lupine. In other larger Pulſe it ſhews likewiſe ſome ob-ſcure Marks of it-ſelf: But in no other Seeds, which I have obſer-ved, though of the greateſt ſize, as of *Apples*, *Plums*, *Nuts*, &c. is there any clear appearance hereof, upon diſſection, ſaving in the *Ra-dicle* and *Plume*; the reaſon of which is partly from its quantity, being in moſt Seeds ſo extraordina-ry little; partly from its Co-lour, which in moſt Seeds, is the ſame with that of the *Parenchyma* it ſelf, and ſo not diſtinguiſhable from it.

Yet in a *Gourd-Seed*, the whole *Seminal Root*, not only its *Main Branches*, but alſo the Sub-diviſions and Inosculations of the lesser ones, are without any dissection, upon the separation of the Lobes, on their contiguous Flats immediatly apparent. See *Fig* 5. And as to the existence of this Seminal Root, what Dissection cannot attain, ocular inspection in hundreds of other Seeds, even the smallest, will demonstrate; as in this *Chapter* shall be seen how.

In the mean time, let us only take notice, that we say every Plant hath its Root, we reckon short; for every Plant hath really two, though not contemporary, yet successive Roots; its Original or *Seminal-Root* within its Seed, and its *Plant-Root*, which the *Radicle* becometh in its growth: the *Parenchyma* of the Seed being in some resemblance, that to the *Seminal Root* at first, which the Mould is to the *Plant-Root* afterwards; and the *Seminal Root* being

ing that to the *Plant-Root*, which the *Plant-Root* is to the *Trunk*. For our better underſtanding whereof, having taken a view of the ſeveral Parts of a Bean, as far as Diſſection conducts; we will next briefly enquire into the uſe of the ſaid Parts, and in what manner they are the Fountain of Vegetation, and concurrent to the being of the future Plant.

The general Cauſe of the growth of a *Bean* or other Seed, is *Fermentation*; that is, the *Bean* lying in the Mould, and a moderate acceſs of ſome moiſture, partly diſſimilar, and partly congenerous, being made, a gentle *Fermentation* thence ariſeth; by which the *Bean* ſwelling, and the *Sap* ſtill encreaſing, and the *Bean* continuing ſtill to ſwell, the work thus proceeds: as is the uſual way of explicating. But that there is ſimply a *Fermentation*, and ſo a ſufficient ſupply of *Sap*, is not e-nough; but that this *Fermentation* and the *Sap* wherein 'tis made, ſhould be under a various Government by divers Parts thereto ſubſervient, is alſo requiſite; and as the various preparation of the *Aliment* in the *Animal*, equally neceſſary; the particular proceſs of the Work according whereto, we find none undertaking to declare.

Let us look upon a *Bean* then, as a piece of Work ſo fram'd and ſet together, as to declare a Deſign for the production of a Plant, which, upon its lying in ſome convenient Soyl, is thus effected. Firſt of all, the *Bean* being enfoulded round in its Coats, the *Sap* wherewith it is fed, muſt of neceſſity paſs through theſe: By which means, it is not only in a proportionate quantity, and by due degrees; but alſo

in

in a purer body; and poſſibly not without ſome Vegetable Tincture, tranſmitted to the *Bean*. Whereas, were the *Bean* naked, the *Sap* muſt needs be, as over-copious, ſo but crude and immature, as not being filtred through ſo fine a Cotton as the Coats be. And as they have the uſe of a *Filtre* to the tranſient *Sap*; ſo of a Veſſel to that which is ſtill depoſited within them; being alike accommodated to the ſecurer *Fermentation* hereof, as Bottles or Barrels are to Beer, or any other *Fermentative Liquor*.

And as the *Fermentation* is promoted by ſome Aperture in the Veſſel; ſo have we the *Foramen* in the upper Coat alſo contrived; that if there ſhould be need of ſome more aiery Particles to excite the *Fermentation*, through this they may obtain their Entry: Or, on the contrary, ſhould there be any ſuch Particles or Steams as might damp the genuine proceeding thereof, through this again they may have eaſie iſſue: being that, as a common Paſport here to the *Sap*, which what we call the Bung-hole of the Barrel, is to the new-tunn'd Liquor. That this *Foramen* is truly permeable even in old ſetting *Beans*, appears upon their being ſoak'd for ſome time in Water: For then taking them out, and cruſhing them a little, many ſmall Bubbles will alternately ariſe and break upon it.

The *Sap* being paſſed through the Coats, it next enters the Body of the *Bean*; yet not indiſcriminately neither; but, being filtred through the *Outer Coat*, and fermented both in the Body and Concave of the *Inner*, is by mediation of the *Cuticle*, again more finely filtr'd, and ſo entereth the

Pa-

Parenchyma it ſelf under a fourth Government.

Through which Part the *Sap* paſſing towards the *Seminal Root*, as through that which is of a more ſpatious content; beſides the benefit it hath of a farther percolation, it will alſo find room enough for a more free and active fermenting and maturation herein. And being moreover, part of the true Body of the *Bean*, and ſo with its proper Seminalities or Tinctures copiouſly repleat; the *Sap* will not only find room, but alſo matter enough, by whoſe Energy its *Fermentation* will ſtill be more advanced.

And the *Sap* being duly prepared here, it next paſſeth into all the Branches of the *Seminal Root*, and ſo under a fifth Government. Wherein how delicate 'tis now become, we may conceive by the proportion betwixt the *Parenchy-*

ma and this *Seminal Root*; ſo much only of the beſt digeſted *Sap* being diſcharged from the whole Stock in that, as this will receive. And this, moreover, as the *Parenchyma*, with its proper Seminalities being endowed; the *Sap*, for the ſupply of the *Radicle*, and of the young Root from thence, is duly prepared therein, and with its higheſt Tincture and Impregnation at laſt enriched.

The *Sap* being thus prepared in the Lobes of the *Bean*, 'tis thence diſcharg'd; and either into the *Plume* or the *Radicle*, muſt forthwith iſſue. And ſince the *Plume* is a dependent on the *Radicle*; the *Sap* therefore ought firſt to be diſpenced to this; which accordingly is ever found to ſhoot forth before the *Plume*, and that ſometimes an inch or two in length. Now becauſe the primitive courſe of the *Sap* into the

Radicle,

Radicle, is thus requisite, therefore by the frame of the Parts of the *Bean* is it made necessary too. For we may observe that the two main Branches of the *Seminal Root* in which the several *Ramifications* in either Lobe are all united, commit not themselves into the *Seminal Trunk* of the *Plume*, nor yet so as to stand at right Angles with them, and with equal respect towards them both; but being producted through part of the *Parenchyma* of the *Radicle*, are at last united therein to the main Trunk, and make acute Angles therewith; as may be seen by *Fig.* 2. Now the *Sap* being brought as far as the *Seminal Root* in either Lobe, and according to the conduct thereof continuing still to move, it must needs immediately issue into the same part whereinto the main Branches themselves do, that is, into the *Radicle*. By which *Sap*, thus

thus bringing the ſeveral Tinctures of the parts aforeſaid with it, being now fed; it is no longer a meer *Radicle*, but is made alſo *Seminal*, and ſo becomes a perfect Root.

The *Radicle* being thus impregnate and ſhot into a Root, 'tis now time for the *Plume* to rouze out of its Cloyſters, and germinate too: In order whereto, 'tis now fed from the Root with laudable and ſufficient Aliment. For as the Supplies and motion of the *Sap* were firſt made from the Lobes towards the Root, ſo the Root being well ſhot into the Moulds, and now receiving a new and more copious *Sap* from theſe; the motion hereof muſt needs be ſtronger, and by degrees revert the primitive *Sap*, and ſo move in a contrary courſe, *ſc.* from the *Root* towards the *Plume*; and, by the continuation of the *Seminal*

Trunk,

Trunk, is directly conducted thereinto; by which, being fed, it gradually enlarges and displayes it self.

The course of the *Sap* thus turned, it issues, I say, in a direct Line from the *Root* into the *Plume*, but collaterally into the Lobes also; *sc.* by those two aforesaid Branches which are obliquely transmitted from the *Radicle* into either Lobe. By which Branches the said *Sap* being disbursed back into all the *Seminal Root*, and from thence likewise into the *Parenchyma* of the Lobes; they are both thus fed, and for some time augmenting themselves, really grow; as in *Lupines* is evident.

Yet is not this common to all Seeds; some rot under-ground, as *Corn*; being of a laxer and less Oleous substance, differing herein from most other Seeds; and being not divided into Lobes, but one

entire

entire thick Body. And ſome, although they continue firm, yet riſe not as the great *Garden-Bean*; in which therefore it is obſervable, that the two Main Branches of the Lobes in compariſon with that which ruus into the *Plume*, are but mean, and ſo inſufficient to the feeding and vegetation of the Lobes; the *Plume*, on the contrary, growing ſo luſty, as to mount up without them.

Excepting a few of theſe two kinds, all other Seeds whatſoever, (which I have obſerved) beſides that they continue firm, upon the Vegetation of the *Plume*, mount alſo upwards, and advance above the Mould together with it; as all Seeds which ſpring up with diſſimilar Leaves; the two (for the moſt part two) diſſimilar Leaves, being the very Lobes of the Seed divided, expanded, and thus advanced.

The Impediments of our apprehenſion

hension hereof are the Colour, Size and Shape of the dissimilar Leaves. Notwithstanding, that they are nothing else but the main body of the Seed, how I came first to phansie, and afterwards to know it, was thus: First, I observed in general that the dissimilar Leaves were never jagg'd, but even edg'd: And seeing the even verges of the Lobes of the Seed hereto respondent, I was apt to think, that those which were so like, might prove the same. Next descending to particular Seeds, I observed first of the *Lupine*; that as to its Colour, upon its advance above the Mould, it ever changed into a perfect Green. And why might not the same by parity of Reason be inferr'd of other Seeds? That, as to its size, it grew but little bigger than when first set. Whence, as I discern'd (the Augmentation being but

but little) we here had only the two Lobes: So, (as ſome augmentation there was) I inferr'd the like might be, and that, in farther degrees, in other Seeds.

Next, of the *Cucumber*-Seed, That, as to its Colour, often appearing above ground in its Primitive white, from white it turns to yellow, and from yellow to green, the proper colour of a Leaf: That, as to its ſize, though at its firſt ariſe, the Lobes were little bigger than upon ſetting; yet afterwards as they chang'd their Colour, ſo their Dimenſions alſo, growing to a three-four-five-fold amplitude above their primitive ſize: But whereas the Lobes of the Seed are in proportion, narrow, ſhort and thick, how then come the diſſimilar Leaves to be ſo exceeding broad, or long and thin? The Queſtion anſwers it ſelf: For the diſſimilar Leaves, for that very reaſon are ſo thin, becauſe ſo very broad or long; as we ſee many things, how much they are extended in length or breadth, ſo much they loſe in depth, or grow more thin; which is that which here befalls the now effoliated Lobes. For being once diſimpriſoned from their Coats, and the courſe of the Sap into them now more and more encreaſed, they muſt needs very conſiderably amplifie themſelves; and from the manner wherein the *Seminal Root* is branched in them, that amplification cannot be in thickneſs, but in length or breadth: In both which, in ſome diſſimilar Leaves 'tis very remarkable; eſpecially in length, as in thoſe of *Lettice*, *Thorn-Apple*, and others; whoſe Seeds, although very ſmall, yet the Lobes of thoſe Seeds growing up into Diſſimilar Leaves, are extended an Inch, and ſometimes more,

more, in length; though he that ſhall attempt to get a clear ſight of the Lobes of *Thorn-Apple*, and ſome others, by Diſſection, will find it no eaſie Task; yet is that which may be obtained. From all which, and the obſervation of other Seeds, I at laſt found, that the diſſimilar Leaves of a young Plant, are nothing elſe but the Lobes or *main Body* of its Seed: So that as the Lobes did at firſt feed and impregnate the *Radicle* into a *perfect Root*; ſo the *Root* being perfected, doth again feed, and by degrees amplifie each Lobe into a perfect Leaf.

The Original of the diſſimilar Leaves thus known, we underſtand, why ſome Plants have none; becauſe the Seed either riſeth not, as *Garden-Beans*, *Corn*, &c. Or upon riſing, the Lobes are little alter'd, as *Lupines*, *Peaſe*, &c. Why, though the proper

Leaves

Leaves are often indented round; the dissimilar, like the Lobes, are even-edg'd. Why, though the proper Leaves are often hairy, yet these are ever smooth. Why some have more dissimilar Leaves than two, as *Cresses*, which have six, as the Ingenious Mr. *Sharrock* also observes; the reason whereof is, because the *Main Body* is not divided into two, but six, distinct Lobes, as I have often counted. Why *Radishes* seem at first to have four, which yet after appear plainly two; because the Lobes of the Seed have both a little Indenture, and are both plaited, one over the other. To which we might add,

The use of the dissimilar Leaves is, first, for the protection of the *Plume*; which being but young, and so but soft and tender, is provided with these, as a double Guard, one on either side of it.

For this reaſon it is, that the *Plume* in Corn is truſſed up within a membranous Sheath; and that of a *Bean*, cooped up betwixt a pair of *Surfoyls*; but where the Lobes riſe, there the *Plume* hath neither of them, being both needleſs.

Again, that ſince the *Plume*, being yet tender, may be injur'd not only by the Air, but alſo for want of Sap, the ſupplies from the Root being yet but ſlow and ſparing; that the ſaid *Plume* therefore, by the diſſimilar Leaves, may have the advantage likewiſe of ſome refreſhment from Dew or Rain. For theſe having their Baſis a little beneath that of the *Plume*, and expanding themſelves on all ſides of it, they often ſtand after Rain, like a Veſſel of Water, continually ſoaking and ſuppling it, leſt its new acceſs into the Ayr, ſhould ſhrivel it.

Moreover, that ſince the diſſi-

milar

milar Leaves by their Basis intercept the *Root* and *Plume*, the greater and grosser part of the Sap may be by the way deposited into those; and so the purest proceed into the yet but young and delicate *Plume*, as its fittest Aliment.

Lastly, we have here a demonstration of the being of the *Seminal Root*; which since through the colour or smalness of the Seed, it could not by dissection be observ'd, except in some few; Nature hath here provided us a way of viewing it in the now effoliated Lobes, not of one or two Seeds, but of hundreds; the *Seminal Root* visibly branching it self towards the Cone and Verges of the said Lobes, or now dissimilar Leaves.

CHAP. II.

Of the Root.

HAving examin'd and pursu'd the Degrees of *Vegetation* in the *Seed*, we find its two Lobes have here their utmost period; and, that having conveyed their Seminalities into the *Radicle* and into the *Plume*; these therefore as the Root and Trunk of the Plant still survive: Of these in their order we next proceed to speak; and first, of the *Root*: whereof, as well as of the *Seed*, we must by Dissection inform our selves.

In Dissection of a *Root* then, we shall

ſhall find it with the *Radicle*, as the Parts of an old man with thoſe of a *Fœtus*, ſubſtantially one. The firſt Part occurring is its Skin, the Original whereof is from the Seed: For that extream thin Cuticle which is ſpred over the Lobes of the Seed, and from thence over the *Radicle*, upon the ſhooting of the *Radicle* into a Root, is co-extended, and becomes its Skin.

The next Part is the *Cortical Body*; the Original whereof likewiſe is from the Seed; or the *Parenchyma*, which is there common both to the Lobes and *Radicle*, being by Vegetation augmented and prolonged into the *Root*, is here the *Cortical Body*, or that which is ſometimes called the *Barque*.

The Contexture of this *Cortical Body* may be well illuſtrated by that of a *Sponge*, being a Body Porous, Dilative, and Pliable.

Its Pores, as they are innumerable, ſo extream ſmall. Theſe Pores are not only ſuſceptive of ſo much Moiſture as to fill, but alſo to enlarge themſelves, and ſo to dilate the *Cortical Body* wherein they are; which by the ſhriv'ling in thereof, by being expos'd to the Air, is alſo ſeen. In which dilatation many of its Parts becoming more lax and diſtant, and none of them ſuffering a ſolution of their continuity; 'tis a Body alſo ſufficiently pliable; or, a moſt exquiſitely fine-wrought Sponge.

The Extention of theſe Pores is much alike both by their length and breadth of the Root; which from the ſhrinking up of the *Cortical Body*, in a piece of a cut Root, by the ſame dimenſions, is argu'd.

The proportions of this *Cortical Body* are various: If thin, 'tis called a *Barque*; & thought to ſerve

to

to no other end, than what is usually aſcrib'd to it as a *Barque*; which is a narrow conceit: If a Bulky Body in compariſon with that within it, as in the young Roots of *Cychory*, *Aſparagus*, &c. 'tis here, becauſe the faireſt, therefore taken for the prime Part; which, though, as to Medicinal uſe, it is; yet, as to the private uſe of the Plant, not ſo. The Colour hereof, though it be originally white, yet in the continued growth of the Root, divers Tinctures, as yellow in *Dock*, red in *Biſtort*, are thereinto introduced.

Next within this Part ſtands the *Lignous Body*; the Original whereof, as of the two former, is from the Seed; or, the *Seminal Roots* of both the Lobes, being united in the *Radicle*, and with its *Parenchyma* co-extended, is here in the Root the *Lignous Body*.

The Contexture hereof is, in many of its parts, much more cloſe than that of the *Cortical*; and their Pores very different: For whereas thoſe of the *Cortical* are infinitely numerous, theſe of the *Lignous* are in compariſon, nothing ſo. But theſe, although fewer, yet are they many of them more open, fair, and viſible: as in a very thin Slice cut athwart the young Root of a Tree, and held up againſt the Light, is apparent: Yet not in all equally, in *Coran*-Tree, in *Goosberry*-Tree, *&c.* leſs; in *Oak*, *Plums*, and eſpecially *Damaſcens*, more; in *Elder*, *Vines*, &c. more conſpicuous. And as they are different in number and ſize, ſo alſo (whereon the numerouſneſs of the Pores of the *Cortical Body* principally depends) in their ſhape. For whereas thoſe of the *Cortical Body* are extended much alike both by the length and breadth of the

Root;

Root; theſe of the *Lignous* are only by the length; which, eſpecially in *Vines* and ſome other *Roots*, is evident. Of theſe Pores, 'tis alſo obſervable, that although in *all* places of the *Root* they are viſible, yet moſt fair and open about the *Fibrous Extremities* of ſome *Roots* (and in many *Roots* higher) where there is no Pith. Theſe Pores, as they ſhew in young *Roots* of Trees, ſee in *Fig.* 6, & 7.

This *Lignous Body* lieth with all its Parts, ſo far as they are viſible, in a Circle or Ring; yet are there divers extream ſmall Fibres thence ſhooting, uſually mixed with the *Cortical Body*; and by the ſomewhat different colour of the ſaid *Cortical Body* where they ſtand, may be noted theſe Fibres; the *Cortical Body* and *Skin* all together, properly make the *Barque*.

The proportion betwixt this *Lignous Body* and the *Cortical*, is various,

various, as was ſaid ; yet in this, conſtant, *ſc.* that in the fibrous,and ſmaller Parts of the *Root*, the *Lignous Body* is not only in compaſs, but in quantity the leſs ; running like a ſlender Wyer or Nerve through the other ſurrounding it. They ſtand both together pyramidally, which is moſt common to *Infant-Roots*, but alſo to many other.

The next Part obſervable in the *Root*, is the *Inſertment.* The exiſtence hereof, ſo far as we can yet obſerve, is ſometimes in the *Radicle* of the Seed it ſelf; I cannot ſay alwayes. As to its ſubſtantial nature, we are more certain ; that it is the ſame with that of the *Parenchyma* of the *Radicle* ; being alwayes at leaſt augmented, and ſo, in part, originated from the *Cortical Body*, and ſo, at ſecond hand, from the ſaid *Parenchyma* : For in diſſecting a *Root*, we find, that

that the *Cortical Body* doth not only environ the *Lignous*, but is alſo wedg'd, and in many pieces inſerted into it; and that the ſaid inſerted pieces make not a meer Indenture, but tranſmit and ſhoot themſelves quite through as far as the Pith; which in a thin Slice cut athwart the *Root*, as ſo many lines drawn from the Circumference towards the Center, ſhew themſelves. See *Fig*. 6, & 7.

The Pores of the *Inſertment* are ſometimes, at leaſt, extended ſomewhat more by the breadth of the *Root*, as about the top of the *Root* of *Borage* may be ſeen; and are thus different from thoſe of the *Cortical Body*, which are extended by the length and breadth much alike; and from thoſe of the *Lignous*, being only by its length.

The number and ſize of theſe Inſertions are various. In *Hawthorn*, and ſome others, and eſpecially

cially *Willows*, they are moſt extream ſmall; in *Cherries* and *Plums* they are large; and in *Damaſcens* eſpecially, very fairly apparent. In the *Roots* of ſmall *Plants* they are generally more eaſily diſcoverable; which may lead to the obſervation of them in all.

Theſe Inſertions, although they are continuous through both the length and breadth of the *Ro t*; yet not ſo in all Parts, but by the ſeveral ſhootings of the *Lignous Body* are frequently intercepted. For of the *Lignous Body* it is (here beſt) obſervable, That its ſeveral ſhootings, betwixt which the *Cortical* is inſerted, are not throughout the *Root* wholly diſtinct; but that all along being enarch'd, the *Lignous Body*, both in length and breadth, is thus diſpoſed into Braces or Oſculations. Betwixt theſe ſeveral ſhootings of the *Lignous Body* thus oſculated, the *Cor-*

tical

tical ſhooting, and being alſo oſculated anſwerably Brace for Brace, that which I call the *Inſertment* is fram'd thereof.

Theſe Oſculations are ſo made, that the Pores of the *Lignous Body*, I think, notwithſtanding, ſeldom run one into another; but, for the moſt part, ſtill keep diſtinct; in the ſame manner as ſome of the Nerves, though they meet, and for ſome ſpace are aſſociated together, yet 'tis moſt probable that none of their Fibres are truly inoſculated here, but only in the Plexures.

Theſe Oſculations of the *Lignous Body*, and ſo the interception of the Inſertions of the *Cortical*, are not to be obſerv'd by the traverſe cut of the *Root*, but by taking off the *Barque*, or the *Cortical Body*. In the Roots of Trees, they are generally obſcure; but in Plants, often more diſtinctly apparent;

rent; and especially in a *Turnep*: the appearance whereof, the *Cortical Body* being stripp'd off, is as a piece of close-wrought Network, fill'd up with the Insertions of the said *Cortical Body*. See *Fig*. 8.

The next and last distinct Part of the *Root* is the *Pith*. The substantial nature thereof, is, as was said of the *Insertment*, the same likewise with that of the *Parenchyma* of the Seed. And according to the best observation we have yet made, 'tis sometimes existent in its *Radicle*; in which, the two main Branches of the Lobes both meeting, and being osculated together, are thus dispos'd into one round Trunk, and so environing part of the *Parenchyma*, make thereof a *Pith*; as in either the *Radicle*, or the young *Root* of the great *Bean* or *Lupine*, may, I think, be well seen.

But many times the Original

hereof is immediately from the *Cortical Body*. For in diſſection of divers *Roots* both of Trees and Plants, as of *Barberry* or *Mallows*, it is obſervable, that the *Cortical Body* and *Pith* are both of them participant of the ſame Colour; in the *Barberry* both of them tinged yellow, and in *Mallows* green. In cutting the ſmaller Parts of the Roots of many Plants, as of *Borage*, *Mallows*, *Parſley*, *Columbine*, &c. 'tis alſo evident, that the *Lignous Body* is not there in the leaſt Concave, but ſtandeth perfectly in the Center; and that the Inſertions being gradually multiplied afterwards, the *Pith* at length, towards the thicker parts of the *Root*, ſhews and enlarges it ſelf. Whence it appears, that in all ſuch *Roots*, the *Pith* is not only of the ſame ſubſtantial nature, and by the Inſertions doth communicate with the *Cortical Body*; and that it it is alſo more or leſs augmented by it; which is true of the *Pith* of all *Roots*; but is moreover, by mediation of the ſaid Inſertions, wholly originated from it. The various appearances of the *Inſertions* and *Pith* from the Fibrous Parts to the top of the *Root*, ſee in *Fig.* 9, 10, 11, 12, 13, 14. The Pores of the *Lignous Body*, entire in the ſaid Fibrous Parts, are beſt ſeen when they have lain by a night dry, after cutting.

A farther evidence hereof are the Proportions betwixt the *Cortical Body* and *Pith*. For as about the inferiour Parts of the *Root*, where the *Pith* is ſmall, the *Cortical Body* is proportionably great; ſo about the top, where the *Pith* is enlarged, the *Cortical Body* groweth proportionably leſs, *&c.* becauſe by its Inſertions, 'tis gradually beſtowed into the *Pith*. Likewiſe the peculiar frame of ſome

Roots,

Roots, wherein besides the *Pith*, the *Lignous Body* being divided into a double Ring, there is also a thick Ring, of a white and soft substance, stands betwixt them; and is nothing else but the Insertions of the *Cortical Body* collected into the said Ring; but, towards the top of the Root, being inserted again, thus maketh a large and ample *Pith*; as in *Fennel*-Roots is seen.

The Pores of the *Pith*, as those of the *Cortical Body*, are extended both by the breadth and length of the *Root*, much alike; yet are they more or less of a greater size than those of the *Cortical Body*.

The Proportions of the *Pith*, are various; in Trees, but small; in Plants generally, very fair; in some making by far the greatest part of the *Root*; as in a *Turnep*: By reason of the wide circumference whereof, and so the finer

Concoction and Aſſimilation of its Sap; that part which in moſt old Trunks is a dry and harſh *Pith*, here proves a tender pleaſant meat. The parts of a *Turnep* in the travers cut ſee in *Fig.* 8.

In the Roots of very many Plants, as *Turneps*, *Carrots*, &c. the *Lignous Body*, beſides its main utmoſt Ring, hath divers of its oſculated Fibres diſperſed throughout the Body of the *Pith*; ſometimes all alike, and ſometimes more eſpecially in, or near, its Center; which Fibres, as they run towards the top of the *Root*, ſtill declining the Center, at laſt collaterally ſtrike into its Circumference; either all of them, or ſome few, keeping the Center ſtill; of theſe principally the *Lignous Body* of the Trunk is often originated.

Theſe Fibres, although they are ſo exceeding-ſlender, yet in ſome

Roots,

Roots, as in that of *Flower-de-liz*, they are viſibly concave, each of them, in their ſeveral Cavities alſo emboſoming a very ſmall *Pith*; the ſight whereof, the Root being cut traverſe, and laid in a Window for a day or two to dry, may without Glaſſes be obtain'd. And this is the general account of the *Root*; the declaration of the manner of its growth, with the uſe and ſervice of its ſeveral parts, we ſhall next endeavour.

We ſay then, that the *Radicle* being impregnate, and ſhot into the Moulds, the contiguous moiſture, by the *Cortical Body*, being a Body laxe and Spongy, is eaſily admitted: Yet not all indiſcriminately, but that which is more adapt to paſs through the ſurrounding Cuticle. Which tranſient Sap, though it thus becomes fine, yet is not ſimple; but a mixture of Particles, both in reſpect of thoſe originally in the Root, and amongst themselves, somewhat heterogeneous. And being lodg'd in the *Cortical Body* moderately laxe, and of a Circular form; the effect will be an easie Fermentation. The *Sap* fermenting, a separation of Parts will follow; some whereof will be impacted to the Circumference of the *Cortical Body*, whence the Cuticle becomes a Skin; as we see in the growing of the Coats of Cheeses, of the Skin over divers Liquors, and the like. Whereupon the *Sap* passing into the *Cortical Body*, through this, as through a *Manica Hippocratis*, is still more finely filtred. With which *Sap*, the *Cortical Body* being dilated as far as its *Tone*, without a solution of Continuity will bear; and the supply of the *Sap* still renew'd; and the purest part, as most apt and ready, recedes, with its due Tinctures, from the said

Cortical

Cortical Body, to the *Lignous*. Which *Lignous Body* likewiſe ſuper-inducing its own proper Tinctures into the ſaid *Sap*; 'tis now to its higheſt preparation wrought up, and becomes (as they ſpeak of that of an Animal) the Vegetative *Ros* or *Cambium*: the nobleſt part whereof is at laſt coagulated in, and aſſimilated to the like ſubſtance with the ſaid *Lignous Body*. The remainder, though not united to it, yet tinctur'd therein, thus retreats, that is, by the continual appulſe of the *Sap*, is in part carried off into the *Cortical Body* back again, the *Sap* whereof it now tinctures into good Aliment: So that whereas before the *Cortical Body* was only relaxed in its Parts, and ſo dilated; 'tis now increas'd in real quantity or number of parts, and ſo is truly nouriſh'd. And the *Cortical Body* being ſaturate with ſo much of this Vital *Sap* as ſerves it ſelf; and the ſecond Remainders diſcharged thence to the Skin; this alſo is nouriſh'd and augmented therewith. So that as in an *Animal Body* there is no inſtauration or growth of Parts made by the Bloud only, but the *Nervous Liquor* is alſo thereunto aſſiſtant; ſo is it here: the *Sap* prepared in the *Cortical Body*, is as the Arterious; and that part thereof prepared by the *Lignous*, is as the *Nervous Liquor*; which partly becoming Nutriment to it ſelf, and partly being diſcharged back into the *Cortical Body*, and diffuſing its Tincture through the *Sap* there, that to the ſaid *Cortical Body* and *Skin*, becomes alſo true Nutriment, and ſo they all now grow.

In which growth, a proportion in length and breadth is requiſite: which being rated by the benefit of the Plant, both for firm ſtanding and ſufficient Sap, muſt therefore

fore principally be in length. And becauſe it is thus requiſite, therefore by the conſtitution of one of its Parts, *ſc.* the *Lignous Body*, it is alſo made neceſſary. For the Pores hereof, in that they are all extended by its length, the *Sap* alſo according to the frame and ſite of the ſaid Pores will principally move; and that way as its *Sap* moves, the ſame way will the generation of its Parts alſo proceed; *ſc.* by its length. And the *Lignous Body* firſt (that is, by a priority cauſal) moving in length it ſelf; the *Cortical* alſo moves therewith. For that which is nouriſh'd, is extended; but whatever is extended, is mov'd; that therefore which is nouriſh'd, is mov'd: The *Lignous Body* then being firſt nouriſh'd, 'tis likewiſe firſt mov'd, and ſo becomes and carries in it the Principle of all Vegetative motion in the *Cortical*; and ſo they both move in length.

Yet as the *Lignous Body* is the Principle of Motion in the *Cortical*; ſo the *Cortical* is the Moderator of that in the *Lignous*: As in Animal Motions, the Principle is from the Nerves; yet being one given to the Muſcle or Limb, and that moving proportionably to its ſtructure, the Nerves alſo are carried in the ſame motion with it. We ſuppoſe therefore, that as the principal motion of the *Lignous Body* is in length, ſo is its proper tendency alſo to aſcend: But being much exceeded both in Compaſs and Quantity by the *Cortical*, as in the ſmaller parts of the *Root* it is; it muſt needs therefore be over-born and governed by it; and ſo, though not loſe its motion, yet make it that way wherein the *Cortical Body* may be more obedient to it; which will be by deſcent:

ſcent: Yet both of them being ſufficiently pliable, they are thus capable, where the Soyl may oppoſe a direct deſcent, there to divert any way where it is more penetrable, and ſo to deſcend obliquely. For the ſame reaſon it may alſo be, that though you ſet a *Bean* with the *Radicle* upward; yet the *Radicle*, as it ſhoots, declining alſo gradually, is thus arch'd in form of an Hook, and ſo at laſt deſcends. For every declination from a perpendicular Line, is a mixed motion betwixt Aſcent and Deſcent; as that of the *Radicle* alſo is, and ſo ſeeming to be dependent upon the two contrary Tendencies of the *Lignous* and *Cortical Bodies*. What may be the cauſe of thoſe Tendencies (being moſt probably external, and perhaps ſomething of a *Magnetiſme*) is beſides my Task here to enquire.

Now although the *Lignous Body*, by

by the poſition and ſhape of its Pores, principally groweth in length; yet will it in ſome degree likewiſe in breadth: For it cannot be ſuppoſed that the pureſt *Sap* is all received into the ſaid Pores; but that part thereof likewiſe, ſtaying about its *Superficial parts*, is there tinctur'd and agglutinated to them. And becauſe theſe Pores are prolonged by its length; therefore is it much more laxe and eaſily diviſible that way; as in ſlitting a Stick, or cleaving of Timber, and in cutting and hewing them athwart is alſo ſeen. Whence it comes to paſs, that in ſhooting from the Center towards the Circumference, and there finding more room, its ſaid original Laxity doth eaſily in divers places now become greater, aud at length in open Partments plainly viſible. Betwixt which Partments, the *cortical Body*, being bound in on the one hand,

hand, by the ſurrounding Skin and Moulds, and preſſed upon by the *Lignous* on the other, muſt needs inſert it ſelf, and ſo move contrary to it, from the Circumference towards the Center: where the ſaid contrary motions continued as begun, they at laſt meet, unite, and either make or augment the *Pith*. And thus the *Root* is fram'd, and the Skin, the *cortical* and *Lignous Bodies*, ſo as is ſaid, hereunto concurrent. We ſhall next ſhew the uſe of the two other Parts, *ſc.* the *Inſertment* and *Pith*; and firſt of the *Pith*,

One true uſe of the *Pith* is for the better Advancement of the *Sap*, whereof we ſhall ſpeak in the next Chapter. The uſe we here obſerve is for the quicker and higher Fermentation of the *Sap*: For although the Fermentation made in the *Cortical Body* was well ſub-

ſubſervient to the firſt Vegetations, yet thoſe more perfect ones in the *Trunk* which after follow, require a Body more adapted to it, and that is the *Pith*; which is ſo neceſſary, as not to be only common to, but conſiderably large in the *Roots* of moſt Plants; if not in their inferiour parts, yet at their tops. Where though either deriv'd or amplify'd from the *Cortical Body*, yet being by its Inſertions only, we may therefore ſuppoſe, as thoſe, ſo this, to be more finely conſtituted. And being alſo from its coarctation, while inſerted, now free; all its Pores, upon the ſupply of the *Sap*, will more or leſs be amplified: Upon which accounts, the *Sap* thereinto received, will be more pure, and its fermentation therein more active. And as the *Pith* is ſuperiour to the *Cortical Body* by its Conſtitution, ſo by its Place. For as it thus ſtands central,

tral, it hath the *Lignous Body* surrounding it. Now as the Skin is the Fence of the *Cortical Body*, and that of the *Lignous* ; so is the *Lignous* again a far more prehemminent one unto the *Pith*; the *Sap* being here a brisk Liquor, tunn'd up as in a wooden Cask.

And as the *Pith* subserves the higher Fermentation of the *Sap*; so do the Insertions its purer Distribution; that separation which the parts of the *Sap*, by being fermented in the *Pith*, were dispos'd for; being, upon its entrance into the Insertions, now made: So that as the Skin is a Filtre to the *Cortical Body*, so are the Insertions a more preheminent one to the *Lignous*; and as they subserve the purer, so the freer and sufficient distribution of the *Sap*: For the *Root* enlarging, and so the *Lignous Body* growing thicker, although the *Cortical* and the *Pith* might supply

ply *Sap* ſufficient to the nutrition of its Parts next adjacent to them; yet thoſe more inward, muſt needs be ſcanted of their *Aliment*; and ſo, if not quite ſtarv'd, yet be uncapable of equal growth: Whereas the *Lignous Body* being through its whole breadth frequently diſparted, and the *Cortical Body* inſerted through it; the *Sap* by thoſe Inſertions, as the Blood by the diſſeminations of the *Arteries*, is freely and ſufficiently convey'd to its intimate Parts, even thoſe which from either the *cortical Body* or the *Pith* are moſt remote. Laſtly, as the conſequent hereof, they are thus aſſiſtant to the Latitudinal growth of the *Root*; as the *Lignous Body* to its growth in length; ſo theſe Inſertions of the *Cortical*, to its better growth in breadth.

Having thus ſeen the ſolitary uſes of the Several Parts of the *Root*, we ſhall laſtly propound our

Con-

Conjectures of that Design whereto they are all together concurrent, and that is the Circulation of the *Sap*: For the *Sap* moving through the *cortical Body*, towards the *Pith*, through the Insertions thereinto, obtains a pass: Which passage, the superiour Insertions will not favour; because the *Pith* standing in the same height with them, is there large, the fermenting and course of the *Sap* quick, and so its opposition strong. But through the inferiour it will much more easily enter; because there, through the smalness of the *Pith*, the opposition is little, and through the shortness of the Insertions, the way more open. So that though the *Sap* may meet with some opposition even here, yet here meeting with the least, here it will bestow it self (feeding the *Lignous Body* in its passage) into the *Pith*. Into which fresh *Sap* still entring, this, being yet but crude, will ſubſide: that firſt receiv'd and ſo become a Liquor higher wrought, will more eaſily mount upwards; and moving in the *Pith*, as in the *Arteria magna*, in equal altitude with the more ſuperiour Inſertions; the moſt volatile parts of all will ſtill continue their direct aſcent towards the *Trunk*. But thoſe of a middle nature, and, as not apt to aſcend, ſo being lighter than thoſe beneath them, not to deſcend neither; they will tend from the *Pith* towards the Inſertions in a motion betwixt both; through which Inſertions (feeding the *Lignous Body* in its paſſage) it is, by the next ſubſequent *Sap*, diſcharged off into the *cortical Body*, as into the *Vena cava*, back again. Wherein, being ſtill purſu'd by freſh *Sap* from the Center, and more occurring from the Circumference, towards the inferiour Inſertions

ertions it thus deſcends; through which, together with part of the *Sap* afreſh imbib'd from the Mould, it re-enters the *Pith*. From whence, into the *Cortical Body*, and from thence into the *Pith*, the cruder part thereof reciprocally is diſburs'd; while the moſt Volatile, not needing the help of a Circulation, more directly aſcendeth towards the *Trunk*.

CHAP. III.

Of the Trunk.

HAving thus declar'd the degrees of *Vegetation* in the *Root* ; the continuance hereof in the *Trunk* ſhall next be ſhew'd: in order to which, the Parts whereof this likewiſe is compounded, we ſhall firſt obſerve.

That which without diſſection ſhews it ſelf, is the *Coarcture*: I cannot ſay of the *Root*, nor of the *Trunk*; but what I chuſe here to mention, as ſtanding betwixt them, and ſo being common to them both; all their Parts being here bound in cloſer together, as in in the tops of the grown Roots of very many Plants, is apparent.

Of the Parts of the *Trunk*, the first occurring is its *Skin*: The Formation whereof, is not from the Air, but in the *Seed*, from whence it is originated; being the production of the Cuticle, there investing the two *Lobes* and *Plume*.

The next Part is the *Cortical Body*; which here in the *Trunk* is no new substantial formation; but, as is that of the *Root*, originated from the *Parenchyma* of the *Seed*; and is only the increase and augmentation thereof. The *Skin*, this *Cortical Body* properly so call'd, and (for the most part) some Fibers of the *Lignous* mixed herewith, all together make the *Barque*.

Next, the *Lignous Body*, which, whether it be visibly divided into many softer Fibres, as in *Fennel*, and most Plants; or that its parts ſtand more compact and cloſe. ſhewing one hard, firm and ſolid piece, as in Trees; it is in all one and the ſame Body; and that not formed originally in the *Trunk*, but in the *Seed*; being nothing elſe but the prolongation of the *Inner Body* diſtributed in the *Lobes* and *Plume* thereof.

Laſtly, The *Inſertions* and *Pith* are here originated likewiſe from the *Plume*, as the ſame in the *Root* from the *Radicle*: So that as to their ſubſtantial Parts, the *Lobes* of the Seed, the *Radicle* and *Plume*, the *Root* and *Trunk* are all one.

Yet ſome things are more fairly obſervable in the *Trunk*. Firſt, the *Latitudinal* ſhootings of the *Lignous Body*, which in *Trunks* of ſeveral years growth, are viſible in ſo many Rings, as is commonly known: For ſeveral young Fibres of the *Lignous Body*, as in the *Root*, ſo here, ſhooting into the

Cortical

Cortical one year, and the ſpaces betwixt them being after fill'd up with more (I think not till) the next, at length they become altogether a firm compact Ring; the perfection of one Ring, and the ground-work of another being thus made concomitantly.

From theſe Annual younger Fibres it is, that although the *Cortical Body* and *Pith* are both of the ſame ſubſtantial nature, and their Pores little different; yet whereas the *Pith*, which the firſt year is green, and of all the Parts the fulleſt of *Sap*, becomes afterwards white and dry; the *Cortical Body*, on the contrary, ſo long as the Tree grows, ever keepeth green and moiſt, *&c.* becauſe the ſaid Fibers annually ſhoot into, and ſo communicate with it.

The Pores likewiſe of the *Lignous Body*, many of them in well-grown Timber, as in Oaken boards, are very conſpicuous, in cutting both lengthwiſe and traverſe; they very ſeldom run one into another, but keep, like ſo many ſeveral Veſſels, all along diſtinct; as by cutting, and ſo following any one of them as far as you pleaſe, for a Foot or half a Yard, or more together, may be obſerv'd.

Theſe greater Pores, though in Wainſcot, Tables, and the like, where they have lain long open, they are but meer Vacuities, and ſo would be thought to contain only *Sap* in the Tree, and afterwards only Air; yet upon a freſh cut, each of them may be ſeen fill'd up with a light and ſpongie Body, which by Glaſſes, and even by the bare eye, appears to be a perfect *Pith*; ſometimes entire, and ſometimes more or leſs broken.

Beſides theſe, there are a leſſer ſort; which, by the help of a *Microſcope*, alſo appear, if not to be fill'd fill'd up with a *Pith*, yet to contain certain light and filmy parts, more or fewer, of a *Pithy* nature within them.

And these are all the Pores the best Glasses, which, (when upon these Enquiries) we had at hand, would shew us. But the Learned and most Ingenious Naturalist Mr. *Hook* sheweth us moreover, besides these, a third, and yet smaller sort; the description whereof I find he hath given us amongst his *Microscopical Observations*. Of these Pores (as a confirmation of what, in the Second Chapter, I have said of the Pores of the *Lignous Body* in general) he also demonstrates; that they are all continuous and prolonged by the length of the *Trunk*, as are the greater ones; the Experiment whereof he imparteth to be, by filling up, suppose in a piece of *Char-coal*, all the said Pores with

Mercury; which appears to paſs quite through them, in that by a very good Glaſs it is viſible in their Orifices at both ends; and without a Glaſs, by the weight of the Coal alone, is alſo manifeſt.

Upon farther Enquiry, I likewiſe find, that the Pores of the *Lignous Body* in the *Trunk* of Plants, which at firſt we only ſuppoſed, by the help of good Glaſſes are very fairly viſible; each Fibre being perforated by 30, 50, 100, or hundreds of Pores. Or what I think is the trueſt notion of them, that each Fibre, though it ſeem to the bare eye to be but one, yet is indeed a great number of Fibres together; every Pore being not meerly a ſpace betwixt the ſeveral pores of the Wood, but the Concave of a Fiber: So that if it be asked, what all that part of a Vegetable, either Plant or Tree, which is properly call'd the woody part,

part; what all that is, I ſuppoſe, that is nothing elſe but a Cluſter of innumerable and moſt extraordinary ſmall Veſſels or concave Fibres. See *Fig.* 15.

Next the Inſertions of the *Cortical Body*, which in the *Trunk* of a Tree ſaw'd athwart, are plainly diſcerned as they run from the Circumference toward the Center; the whole Body of the Tree being viſibly compounded of two diſtinct Subſtances, that of the ſeveral Rings, and that of the Inſertions, running croſs; ſhewing that in ſome reſemblance in a Plain, which the Lines of Latitude and of the Meridian do in a Globe. See *Fig.* 16.

Theſe Inſertions are likewiſe very conſpicuous in Sawing of Trees length-wayes into Boards, and thoſe plain'd, and wrought into Leaves for Tables, Wainſcot, Trenchers, and the like. In all which,

which, as in courſe Trenchers made of *Beech*, and Tables of Oak, there are many parts which have a greater ſmoothneſs than the reſt; and are ſo many inſerted pieces of the *Cortical Body*; which by reaſon of thoſe of the *Lignous*, ſeem to be diſcontinuous, although in the *Trunk* they are extended throughout its Breadth.

Theſe Inſertions, although as is ſaid, of a quite diſtinct ſubſtance from the *Lignous Body*, and ſo no where truly incorporated with it, yet being they are in all parts, the one as the Warp, the other as the Woof, mutually braced and interwoven together, they thus conſtitute one ſtrong and firmly coherent Body.

As the Pores are greater or leſs, ſo are the Inſertions alſo: To the bare eye uſually the greater only are diſcernable: But through an indifferent *Microſcope* there are others

others alſo, much more both numerous and ſmall, diſtinctly apparent. So that, I think, we may obſerve, that as the grand *Pith* of the *Trunk* communicates with, and is augmented by the greater Inſertions; ſo is the *Pith* of each greater Pore originated from the leſs; and thoſe (at leaſt) pithy parts in the Midling Pores, from others ſtill leſs; and ſuppoſe, that the leaſt of all are ſo far intruded into the ſmalleſt Pores, as only juſt to cauſe a kind of roughneſs on their concave ſides, and no more; to what end ſhall be ſaid See *Fig.*17.

In none of all theſe Pores can we obſerve any thing which may have the true nature and uſe of Valves, which is eaſily to admit that, to which they will by no means allow a regreſs. And their non-exiſtence is enough evident, from what in the firſt Chapter we have ſaid of the *Lobes* of the *Seed*: in in whoſe *Seminal Root*, were there any *Valves*, it could not be, that by a contrary courſe of the *Sap*, they ſhould ever grow; which yet, where-ever they turn into Diſſimilar Leaves, they do. Or if we conſider the growth of the *Root*, which oftentimes is upward and downward both at once.

The Inſertions here in the *Trunk*, give us likewiſe a ſight of the poſition of their Pores. For in a plained piece of Oak, as in Wainſcot, Tables, *&c.* beſides the larger Pores of the *Lignous Body*, which run by the length of the *Trunk*; the Tract likewiſe of thoſe of the Inſertions may be obſerved to be made by the breadth, and ſo directly croſs. Nor are they continuous as thoſe of the *Lignous Body*, but very ſhort, as thoſe both of the *Cortical Body* and *Pith*, with which the Inſertions, as to their ſubſtance are congenerous. Yet

they

they all ſtand ſo together, as to be plainly ranked in even Lines or Rows throughout the breadth of the *Trunk*: As the Tract of theſe Pores appear to the naked Eye, ſee in *Fig.* 18. By the beſt Microſcope I have at hand, I can only obſerve the Ranks of the Pores; not the Pores themſelves, ſaving here and there one; wherefore I have not deſcrib'd them.

The Pores of the *Pith* likewiſe being larger here in the *Trunk*, are better obſervable than in the *Root*: the width whereof, in compariſon with their ſides ſo exquiſitely thin, may by an Honey-Comb be groſly exemplified; and is that alſo which the vaſt diſproportion betwixt the Bulk and weight of a dry *Pith* doth enough declare. In the *Trunks* of ſome Plants, they are ſo ample and tranſparent, that in cutting both by the length and breadth of the *Pith*, ſome of them, even to the

the bare eye would ſeem to be conſiderably extended by the length of the ſaid *Pith*; which once I alſo thought they were, and that only the reſt of them were but ſhort and diſcontinuous, and as 'tis ſaid, ſomewhat anſwerable to the Cells of an Honey-Comb. This was the neareſt we could come to them, by conjecture, and the aſſiſtance of the beſt Glaſſes we then had by us, when upon enquiry into the nature of the *Pith*: But that Worthy Perſon newly mentioned Mr. *Hooke* ſheweth us, that the Pores of the *Pith*, particularly of *Elder-Pith*, ſo far as they are viſible, are all alike diſcontinuous; and that the *Pith* is nothing elſe (to uſe his own words) but an heap of Bubbles.

Beſides what this Obſervation informs us of here, it farther confirms what in the ſecond Chapter we have ſaid of the Original of the

Pith

Pith and *Cortical Body*, and of the ſameneſs of both their natures with the *Parenchyma* of the Seed. For, upon farther enquiry with better Glaſſes, I find, that the *Parenchyma* of the *Plume* and *Radicle*, and even of the *Lobes* themſelves, though not ſo apparently, is nothing elſe but a Maſs of Bubbles.

In the *Piths* of many Plants, the greater Pores have ſome of them leſſer ones within them, and ſome of them are divided with croſs Membranes: And betwixt their ſeveral ſides, have, I think, other ſmaller Pores viſibly interjected. However, that they are all permeable, is moſt certain. They ſtand together not indeterminately, but in even Ranks or Trains; as thoſe of the Inſertions by the breadth, ſo theſe by the length of the *Trunk*. And thus far there is a general corresponding betwixt the part of the *Root* and *Trunk*: Yet are there ſome

ſome conſiderable Diſparities be-twixt them; wherein, and how they come to paſs, and to what eſpecial uſe and end, ſhall next be ſaid.

We ſay then, that the *Sap* be-ing in the *Root* by Filtrations, Fermentations (and in what *Roots* needful, perhaps by Circulation alſo) duly prepar'd; the prime part thereof paſſing through the intermediate Coarcture, in due moderation and purity is enter-tain'd at laſt into the *Trunk*. And the *Sap* of the *Trunk* being purer and more volatile, and ſo it ſelf apt to aſcend; the motion of the *Trunk* likewiſe will be more no-ble, receiving a diſpoſition and tendency to aſcend therewith. And what by the *Sap* the *Trunk* is in part diſpos'd to, by the reſpe-ctive poſition and quantity of its Parts it is effectually enabled. For whereas in the *Root* the *Lignous*

Body

Body being in proportion with the *Cortical*, but little, and all lying close within its Center; it must therefore needs be under its controul: on the contrary, being here comparatively of greater quantity, and also more dilated, and having divers of its Branches standing more abroad towards the Circumference, as both in the Leaves and Body of the young *Trunk* and *Plume*, 'is seen; it will in its own tendency to ascend, reduce the *Cortical Body* to a compliance with it.

And the *Trunk* thus standing from under the restraint of the Mould in the open Air, the disposition of its Parts originally different from that of the Parts in the Root will not only be continued, but improved: For by the force and pressure of the *Sap* in its collateral Motion, the *Lignous Body* will now more freely and farther be dilated. And this being dilated, the *Cortical Body* alſo, muſt needs be inſerted; and is therefore in proportion alwayes more or leſs ſmaller here in the *Trunk*, than in the *Root*. And as the *Cortical Body* leſſens, ſo the *Pith* will be enlarged, and by the ſame proportion is here greater, And the *Pith* being enlarged it ſelf, its Pores (the *Lignous Body*, upon its dilatation, as it were tentering and ſtretching out all their ſides) muſt needs likewiſe be enlarged with it, and accordingly are ever greater in the *Pith* of the *Trunk*, than of the *Root*. And the dilatation of the *Lignous Body* ſtill continued, it follows, that whereas the *Pith* deſcendent in the *Root*, is not only in proportion leſs and leſs, but alſo in the ſmaller extremities thereof and ſometimes higher altogether abſent: Contrariwiſe, in the *Trunk* it is not only con-

continued to its top, but alſo there in proportion equally ample with what it is in any other inferiour part.

But although the openneſs of the Ayr permitting be alwayes alike; yet the Energy of the *Sap* effecting, being different; as therefore that doth, the dilatation of the *Trunk* will alſo vary. If that be leſs, ſo is this; as in the *Trunks* of moſt Trees: If that be greater, ſo is this; as in Plants is common; the *Lignous Body* being uſually ſo far dilated, that the *utmoſt Shootings* thereof may eaſily be ſeen to jut out, and adjoyn to the Skin. And if the *Sap* be ſtill of greater energy, it ſo far dilates the *Lignous Body*, as not only to amplifie the *Pith* and all its Pores; but alſo ſo far to ſtretch them out, as to make them tear. Whereupon either running again into the *Cortical Body*, or ſhrinking up towards it, the *Trunk* thus ſometimes becomes an *hollow Stalk*, the *Pith* being wholly, or in part voyded. But generally it keeps entire; and where it doth, the ſame proportion and reſpect to the *Lignous* and *Cortical-Bodies*, as is ſaid. The Conſequences of all which will be, the ſtrength of the *Trunk*, the ſecurity and plenty of the *Sap*, its Fermentation will be quicker, its Diſtribution more effectual, and its Advancement more ſufficient.

Firſt, the erect growth and ſtrength of the *Trunk*; this being by the poſition of its ſeveral parts effected: For beſides the ſlendering of the *Trunk* ſtill towards the top, the Circumferential poſition of the *Lignous Body* likewiſe is, and that eminently hereunto ſubſervient: So that as the *Lignous Body* in the ſmaller part of the *Root* ſtanding Central, we may thence conceive

conceive and ſee théir pliableneſs to any oblique motion ; ſo here, on the contrary, the *Lignous Body* ſtanding wide , it thus becomes the ſtrength of the *Trunk*, and moſt advantageous to its perpendicular growth. We ſee the ſame Deſign in *Bones* and *Feathers* : The ſtrongeſt *Bones*, as thoſe in the Legs, are hollow. Now ſhould we ſuppoſe the ſame *Bone* to be contracted into a Solid Body, although now it would be no heavier , and in that reſpect, as apt for motion; yet would it have far leſs ſtrength, than as it is dilated to a Circumferential poſture. And ſo for *Quills*, which, for the ſame Reaſons, in ſubſerviency to flight, we ſee how exceeding light they are, and yet, in compariſon with the thinneſs of their Body, how very ſtrong: We ſee it not only in Nature, but Art. For hence it is that *Joyners* and *Carpenters* unite and ſet together their Timber-pieces and ſeveral Works oftentimes with double Joynts; which, although they are no thicker than a ſingle one might be made, yet ſtanding at a diſtance, have a greater ſtrength than that could have. And the ſame Architecture will have the ſame uſe in the *Trunks* of Plants, in moſt whereof 'tis very apparent; as for inſtance, in Corn: For Nature deſigning its *Sap* a great Aſcent for its higher maturity, hath given it a tall *Trunk*; but to prevent its ravenous deſpoiling either of the Ear or Soyl; although it be tall, yet are its ſides but thin: and becauſe again, it ſhould grow not only tall and thriftily, bnt for avoiding propping up, ſtrongly too; therefore, as its height is over-proportioned to the thinneſs of its ſides, ſo is its Circumference alſo; being ſo far dilated as to parallel a *Quill* it it ſelf. Beſides the poſition of the *Lignous Body* within the compaſs of a Ring, we ſee ſome ſhootings thereof often ſtanding beyond the Circumference of the ſaid Ring, making ſometimes a triangular, oftner a quadrangular Body of the *Trunk*; to the end, that the Ring being but thin, and not ſelf-ſufficient, theſe, like Splinters to *Bones*, might add ſtrength and ſtability to it.

Next, the ſecurity and plenty of the *Sap*. For ſhould the *Lignous Body*, as it doth in the *Root*, its ſmaller parts, ſtand Central here alſo, and ſo the *Cortical* wholly ſurround it: the greater part of the *Sap* would thus be more immediately expos'd to the Sun and ayr; and being lodg'd in a laxe Body, by them continually be prey'd upon, and as faſt as ſupplied to the *Trunk*, be exhauſted. Whereas the *Pith* ſtanding in the

 Center,

Center, the *Sap* therein being not only moſt remote from the Ayr and Sun, but by the *Barque*, and eſpecially the *Wood*, being alſo ſurrouuded and doubly immur'd, will very ſecurely and copiouſly be convey'd to all the Collateral parts, and (as ſhall be ſaid how) the top of the *Trunk*.

And the *Sap* by the amplitude, and great poroſity of the *Pith* being herein more copious, its Fermentation alſo will be quicker; which we ſee in all Liquors by ſtanding in a greater quantity together, proceeds more kindly: And being tunn'd up within the *Wood*, is at the ſame time not only ſecur'd from loſs, but all extream mutations, the Day being thus not too hot, nor the Night too cold for it.

And the Fermentation hereof being quicker, its motion alſo will be ſtronger, and its diſtribution more

more effectual, not only to the dilatation of the *Trunk*, but likewise the shooting out of the Branches. Whence it is, that in the Bodies of Trees, the *Barque* of it self, though it be sappy, and many Fibres of the *Lignous Body* mixed with it, yet seldom sendeth forth any; and that in Plants, those with the least *Pith* (other advantages not supplying this defect) have the fewest or smallest Branches, or other collateral Growths: and that *Corn*, which hath no *Pith*, hath neither any Branches.

Lastly, the Advancement of the *Sap* will hence also be more ready and sufficient. For the understanding where, and how, we suppose that in all *Trunks* whatsoever there are two parts joyntly hereunto subservient. In some the *Lignous Body* and the *Cortical*, as in older *Trunks*, the *Pith* being either ex-

excluded or dried: But in moſt, principally the *Lignous Body* and *Pith*; as in moſt Annual Growths of Trees; but eſpecially Plants, where the *Cortical Body* is uſually much and often wholly inſerted.

Of the *Lignous* body it is ſo apparent by its Pores, or rather by its Veſſels, that we need no farther evidence. For as to what end are Veſſels but for the conveyance of Liquor? And is that alſo, which upon cutting the young Branch of a Sappy Tree or Plant, by an accurate and ſteady view may be obſerved. But when I ſay the Pores of the *Lignous Body*, I mean principally them of the younger ſhootings, both thoſe which make the new Ring, and thoſe which are mixed with the *Cortical Body* in the *Barque*: that which aſcendeth by the Pores of the older Wood, being probably, becauſe in leſs quan-

quantity, more in form of a Vapour, than a Liquor. Yet that which drenching into the ſides of its Pores, is withall thereunto ſufficient Aliment; as we ſee *Orpine*, *Onions*, &c. only ſtanding in a moyſter Ayr will often grow; And being likewiſe in part ſupplied by the Inſertions from the younger Shoots: But eſpecially, becauſe as it is but little, ſo it ſerveth only for the growth of the ſaid *Elder Wood*, and no more; whereas the more copious Aliment aſcendent by the younger Shoots, ſubſerves not only their own growth, but the generation of others; and is beſides with that in the *Cortical Body* the Fountain of *Perſpirations*, which we know even in Animals are much more abundant than the *Nutritive parts*; and doubtleſs in a *Vegetable* are ſtill much more.

But theſe Pores, although they are a free and open way to the aſcend-

aſcending *Sap*; yet that meer Pores or Veſſels ſhould be able of themſelves to advance the *Sap* with that ſpeed, ſtrength & plenty, and to that height, as is neceſſary, cannot probably be ſuppoſed. It follows then, that herein we muſt grant the *Pith* a joynt ſervice. And why elſe in the ſmaller parts of the *Root*, where the *Pith* is often wanting, are the Pores there greater? Why is the *Pith* in all primitive growths the moſt *Sappy* part, why hath it ſo great a ſtock of *Sap*, if not after due maturation within it ſelf ſtill to be disburſed into the Fibres of the *Lignous Body*? Why are the annual growths of all both Plants and Trees with great Piths, the quickeſt and the longeſt? But how are the Pores of the *Pith* permeable? That they are ſo, both from their being capable of a repletion with *Sap*, and of being again wholly emptied of it, and

again,

again, instead thereof fill'd with Ayr, is as certain as that they are Pores. That they are permeable, by the breadth, appears from the dilatation of the *Lignous Body*, and and from the production of Branches, as hath been, and shall hereafter be said. And how else is there a Communion betwixt this and the *Cortical Body*? That they are so also, by the length, is probable, because by the best *Microscope* we cannot yet obseiv, that they are visibly more open by the breadth, than by the length. And withal are ranked by the length, as those of the Insertions by the breadth of the *Trunk*. But if you set a piece of dry *Elder-Pith* in some tinged Liquor, why then doth it not penetrate the Pores, so as to ascend through the Body of the *Pith*? The plain reason is, because they are all fill'd with Ayr. Whereas the *Pith* in a Vegetating Plant, as its Parts

or

or Pores are still generated, they are at the same time also fill'd with *Sap*; which, as 'tis gradually spent, is still repair'd by more succeeding, and so the Ayr still kept out; as in all primitive growths, and the *Pith* of *Elder* it self: Yet the same *Pith*, by reason of the following Winter, wanting a more copious and quick supply of *Sap*, thus once become, ever after keeps dry. And since in the aforesaid Trial the Liquor only ascends by the sides of the *Pith*, that is of its broken Pores, we should thence by the same reason conclude that they are not penetrable by the breadth neither, and so no way; and then it need not be ask'd what would follow. But certainly the *Sap* in the Pores of the *Pith* is discharged and repaired every moment, as by its shriv'ling up, upon cutting the Plant is evident.

We suppose then, that as the

Sap

Sap aſcendeth into the *Trunk* by the *Lignous Body*, ſo partly alſo by the *Pith*. For a piece of *Cotton* with one end immers'd in ſome tinged Liquor, and with the other erect above, though it will not imbibe the Liquor ſo far as to overrun at the top, yet ſo as to advance towards it, it will; ſo here, the *Pith* being a porous and ſpongy Body, and in its Vegetating ſtate its Pores alſo permeable, as a curious Filtre of Natures own contrivance, it thus advanceth, or as people uſe to ſay, ſucks up the *Sap*. Yet as it is ſeen of the Liquor in the Cotton; ſo likewiſe are we to ſuppoſe it of the *Sap* in the *Pith*; that though it riſeth up for ſome way, yet is their ſome term, beyond which it riſeth not, and towards which the motion of the aſcending *Sap* is more and more broken, weak and ſlow, and ſo the quantity thereof leſs and leſs. But becauſe the *Sap* moveth not only by the length, but breadth of the *Pith*; at the ſame time therefore as it partly aſcendeth by the *Pith*, it is likewiſe in part preſſed into the *Lignous Body* or into its Pores. And ſince the motion of the *Sap* by the breadth of the *Pith* not being far continued, and but collateral, is more prone and eaſie than the perpendicular, or by its length; it therefore follows, that the collateral motion of the *Sap*, at ſuch a height or part of the *Pith*, will be equally ſtrong with the perpendicular at another part, though ſomewhat beneath it; and that where the perpendicular is more broken and weak, the collateral will be leſs; and conſequently where the perpendicular tendency of the *Sap* hath its term, the collateral tendency thereof, and ſo its preſſure into the Pores of the *Lignous Body* will ſtill continue.

Through

Through which, in that they are ſmall, and ſo their ſides almoſt contiguous, the *Sap* as faſt as preſſed into them will eaſily run up; as betwixt the two halves of a Stick firſt ſlit, and then tied ſomewhat looſely together, may alſo any Liquor be obſerved to do. And the ſides of the ſaid Pores being not ſmooth, but by the intruſion of the ſmalleſt inſertions made ſomewhat rough; by that means the higher and more facile aſcent of the *Sap* therein will farther be promoted. By all which Advantages the facility and ſtrength of that aſcent will be continued higher in the ſaid Pores than in the *Pith*. Yet ſince this alſo, as well as that in the *Pith* will have its term; the *Sap*, although got thus far, would yet at laſt be ſtagnant, or at leaſt its aſcent be very ſparing, ſlow and feeble, if not ſome way or other re-inforced. Wherefore, as the

Sap moving by the breadth of the *Pith*, presseth thence into the Pores of the *Lignous Body*; so having well fill'd these, is in part by the same Collateral motion disbursed back into a yet higher Region of the *Pith*. By which partly, and partly by that portion of the *Sap*, which in its perpendicular ascent was before lodged therein; 'tis thus here, as in any inferiour place equally repleat. Whereupon the force and vigour of the perpendicular motion of the *Sap* herein will likewise be renew'd; and so its Collateral motion also, and so its pressure into the Pores of the *Lignous Body*, and consequently its ascent therein; and so by a pressure from these into the *Pith*, and from the *Pith* into these reciprocally carried on, a most ready and copious ascent of the *Sap* will be continued from the bottom to the top, though of the highest *Trunk*.

An Appendix.

Of *Trunk-Rooots* and *Claſpers*.

THe diſtinct Parts whereof theſe are conſtituted, are the ſame with thoſe of the *Trunk*, and but the continuation of them.

Trunk-Roots are of two kinds: Of the one, are thoſe that vegetate by a direct deſcent: The place of their Eruption is ſometimes all along the *Trunk*; as in *Mint*, &c. Sometimes only at its utmoſt point, as in the *Bramble*.

The other ſort are ſuch as neither aſcend nor deſcend, but ſhoot forth at right Angles with the *Trunk*; which therefore, though as to their Office, they are true *Roots*, yet as to their Nature, they are a *Middle thing* betwixt a *Root* and a *Trunk*.

Claspers, though they are but of one kind, yet their nature is double; not a mean betwixt that of the *Root* and that of the *Trunk*, but a compound of both; as in their Circumvolutions, wherein they often mutually aſcend and deſcend, is ſeen.

The uſe of theſe Parts may be obſerved as the *Trunk* mounts, or as it trails. In the mounting of the *Trunk*, they are for ſupport and ſupply: For ſupport, we ſee the *Claſpers* of *Vines*; the Branches whereof being very long, fragile and ſlender, unleſs by their *Claſpers* they were mutually contain'd together, they muſt needs by their own weight, and that of

their

their Fruit, undecently fall, and be alſo liable to frequent breaking. So that the whole care is divided betwixt the Gardener and Nature; the Gardener with his Ligaments of Leather ſecures the main Branches; and Nature with theſe of her own finding, ſecures the leſs. Their Conveniency to which end, is ſeen in their Circumvolutions, a motion not proper to any other Part: As alſo in their toughneſs or ſtrength, though much more ſlender than the Branches whereon they are appendent.

For Supply, we ſee the *Trunk-Roots* of *Ivy*: For mounting very high, and being of a cloſer Conſtitution than that of a *Vine*, the *Sap* could not be ſufficiently ſupplied to the upper Sprouts, unleſs theſe to the *Mother-Root* were joyntly aſſiſtant. Yet ſerve they for ſupport likewiſe; whence they ſhoot out, not as in *Creſſes*, *Brook-*

lime, &c. reciprocally on each ſide, but commonly all in one; that ſo they may be faſtened at the neareſt hand.

In the Trailing of the *Trunk*, they ſerve for ſtabiliment, propagation and ſhade. For ſtabiliment, we ſee the *Claſpers* of *Cucumbers*: For the *Trunk* and *Branches* being long and fragile, the Bruſhes of the Winds would injuriouſly hoiſe them to and fro, to the dammage both of themſelves and their tender Fruit, were they not by theſe Ligaments brought to good Aſſociation and Settlement.

As for this end, ſo for Propagation, we ſee the *Trunk-Roots* of *Camomile*. Whence we have the reaſon of the common obſervation, that it grows better by being trod upon: the Mould, where too laxe, being thus made to lie more conveniently about the ſaid *Trunk-Roots* newly bedded therein; and

is

is that which we ſee alſo effected in Rowling of *Corn*.

For both theſe ends, we ſee the *Trunk-Roots* of *Strawberries*; as alſo for ſhade; for in that we ſee all *Strawberries* delight; and by the trailing of the Plant is well obtain'd: So that as we are wont to tangle the Twigs of Trees together to make an *Arbour Artificial*; the ſame is here done to make a *Natural one*; as likewiſe by the *Claſpers* of *Cucumbers*: For the Branches of the one by the Linking of their *Claſpers*, and of the other by the Tethering of their *Trunk-Roots*, being couched together; their tender fruits thus lie under the Umbrage of a *Bower* made of their own Leaves.

CHAP. IV.

Of the Germen, Branch, and Leaf.

THe Parts of the *Germen* and *Branch*, are the ſame with thoſe of the *Trunk*; the ſame *Skin*, *Cortical* and *Lignous Bodies*, *Inſertment* and *Pith*, hereinto propagated, and diſtinctly obſervable herein.

For upon Enquiry into the Original of a *Branch* or *Germen*, it appears, That it is not from the *Superficies* of the *Trunk*, but ſo deep, as to take with the *Cortical*, the *Lignous Body* into it ſelf; and that not only from its Circumference, but

but (ſo as to take the *Pith* in alſo) ſrom its *Inner* or *Central parts*. Divers whereof may commonly be ſeen to ſhoot out into the *Pith*; from which *Shoots*, the ſurrounding and more ſuperiour *Germens* are originated; in like manner as the *Lignous Body* of the *Trunk* is ſometimes principally from thoſe Fibrous *Shoots* which run along the *Pith* in the *Root*.

The manner wherein uſually the *Germen* and *Branch* are fram'd, is briefly thus: The *Sap* (as is ſaid, *Chap*. 3.) mounting in the *Trunk*, will not only by its length, but by its breadth alſo, through the *Inſertions* partly move. Yet, its Particles being not all alike qualified, in different degrees: Some are more groſs and ſluggiſh; of which we have the formation of a Circle of Wood only, or of an Annual Ring: Others are more brisk; and by theſe we have the

Germen

Germen propagated. For by the vigour of their own motion from the Center, they impreſs an equal tendency on ſome of the inner parts of the *Lignous Body* next adjacent to the *Pith*, to move with them. And ſince the *Lignous Body* is not entire, but frequently diſparted; through theſe diſpartments, the ſaid interiour Parts, upon their Nutrition, actually ſhoot; not only towards the Circumference, ſo as to make part of a Ring, but even beyond it, in order to the production of a *Germen*. And the *Lignous Body* thus moving, and carrying the *Cortical* along with it; they both make a force upon the *Skin*: Yet their motion being moſt even and gradual, that force is ſuch likewiſe; not to cauſe the leaſt breach of its parts, but gently to carry it on with themſelves; and ſo partly by the extenſion of its already exiſtent

ſtent parts, as of thoſe of Gold in drawing of Guilded Wyer; and partly by the accretion of new ones, as in the enlarging of a Bubble above the Surface of the Water, it is extended with them to their utmoſt growth. In which growth, the *Germen* being prolonged, and ſo diſplaying its ſeveral parts, as when a *Proſpective* or *Teleſcope* is drawn out, thus becomes a *Branch*.

The ſame way as the propagation of the Parts of a *Germen* is contriv'd is its due nutrition alſo: For being originated from the inner part of the *Lignous Body*, 'tis nouriſhed with the beſt fermented *Sap* in the *Trunk*, *ſc.* that next adjacent to it in the *Pith*. Beſides, ſince all its Parts, upon their ſhooting forth, divaricate from their perpendicular, to a croſs Line, as theſe and the other grow and thrive together, bind and throng

each

each other into a Knot; through which Knot the *Sap* being ſtrain'd, 'tis thus in due moderation & purity delivered up into the Branch.

And for Knots, they are ſo neceſſary, as to be ſeen not only where collateral Branches put forth; but in ſuch Plants alſo as ſhoot up in one ſingle *Trunk*; as in *Corn*; wherein, as they make for the ſtrength of the *Trunk*; ſo by ſo many percolations as they are Knots, for the tranſmiſſion of the *Sap* more and more refined towards the Ear. So that the two general uſes of Knots are for firmer ſtanding, and finer growth.

Laſtly, as the due Formation and Nutrition of the *Germen* are provided for, ſo is its ſecurity alſo; which both in its poſition upon the *Trunk*, and that of its Parts among themſelves may be obſerved. The poſition of its Parts ſhall be conſidered in ſpeaking of the Leaf.

Leaf. As to its ſtanding in the *trunk*, 'tis alwayes betwixt the *trunk* or *Elder Branch*, and the *Baſis* of the Stalk of the *Leaf*; whereby it is not only guarded from the Injuries of any contingent Violence, but alſo from the more piercing aſſaults of the Cold, ſo long till in time 'tis grown, as larger, ſo more hardy. The manner and uſes of the poſition of every *Germen*, conſidered as after it becomes a *Branch*, hath already been by the Ingenious Mr. *Sharrock* very well obſerved; to whom I refer.

Upon the prolongation of the *Germen* into a *Branch*, its *Leaves* are thus diſplay'd. The Parts whereof are ſubſtantially the ſame with thoſe of a *Branch*: For the Skin of the Leaf is only the ampliation of that of the *Branch*; being partly by the accretion of new, & partly the extention of its already exiſtent parts (dilated as in making

king of *Leaf-Gold*) into its present breadth. The Fibres or Nerves dispersed through the Leaf, are only the Ramifications of the *Branch's* Wood, or *Lignous Body*. The *Parenchyma* of the Leaf which lies betwixt the Nerves, and as in Gentlewomens Needle-works, fills all up, is nothing else but the continuations of the *Cortical Body*, or inner part of the *Barque* from the *Branch* into it self, as in most Plants with a fat Leaf, may easily be seen.

The Fibres of the *Leaf* neither shoot out of the *Branch* nor *Trunk*, nor stand in the *Stalk*, in an even Line; but alwayes in either an Angular or Circular posture, and usually making either a Triangle, or a Semi-Circle, or Cord of a Circle; as in *Cycory*, *Endive*, *Cabbage*, &c. may be observed: And if the Leaf have but one main Nerve, that also is postur'd in a Circular or Lunar Figure;

gure; as in *Mint* and others. The usual number of these Nerves or Fibres is 3, 5, or 7. See the *Figures* from 20, to 29.

The reason of the said Positions of the Fibres in the *Stalk* of the *Leaf*, is for its more erect growth, and greater strength; which, were the position of the said Fibres in an even Line, and so the Stalk it self, as well as the Leaf flat, must needs have been defective; as from what we have said of the Circumferential posture of the *Lignous Body* in the *Trunk*, we may better conceive.

As likewise for the security of its *Sap*: For by this means it is, that the several Fibres, and especially the main or middle Fibre of the Leaf, together with a considerable part of the *Cortical Body*, are so disposed of, as to jut out, not from its upper, but its back, or nether plain. Whence the whole Leaf,

Leaf, reclining backward, becomes a Canopy to them, defending them from thoſe Injuries which from colder Blaſts, or an hotter Sun, they might otherwiſe ſuſtain. So that by a mutual benefit, as theſe give ſuck to all the Leaf, ſo that again protection to theſe.

Theſe Fibres are likewiſe the immediate viſible Cauſe of the ſhape of the Leaf: For if the nethermoſt Fibre or Fibres in the Stalk be in proportion greater, the Leaf is long, as in *Endive*, *Cycory*, and others: If all of a more equal ſize, it ſpreads rounder, as in *Ivy*, *Doves-foot*, *Colts foot*, &c. And although a *Dock-Leaf* be very long, whoſe Fibres notwithſtanding, as they ſtand higher in the Stalk, are diſpoſed into a Circle all of an equal ſize; yet herein a peculiar fibre, ſtanding in the Center betwixt the reſt, and running through the length of the Leaf, may be obſerved.

In

In correſpondence alſo to the ſize and ſhape of theſe Fibres, is the Leaf flat: In that either they are very ſmall, or if larger, yet they never make an entire Circle or Ring; but either half of one, as in *Borage*, or at moſt three parts of one, as in *Mullen*, may be ſeen. For if either they were ſo big, as to contain; or ſo entire, as perfectly to include a *Pith*, the Energy of the *Sap* in that *Pith*, would cauſe the ſaid *Lignous Ring* to ſhoot forth on every ſide, as it doth in the *Root* or *Trunk*: But the ſaid Fibres being not figur'd into an entire Ring, but ſo as to be open; on that hand therefore where open, they cannot ſhoot any thing directly from themſelves, becauſe there they have nothing to ſhoot; and the *Sap* having alſo a free vent through the ſaid opening, againſt that part therefore which is thereunto oppoſite, it can have no force;

and ſo neither-will they ſhoot forth on that hand; and ſo will they conſequently that way only which the force of the *Sap* directs, which is only on the right and left.

The ſeveral Fibres in the Stalk, are all inoſculated in the Leaf, with very many Sub-diviſions; according as theſe Fibres are inoſculated near, or at, or ſhoot directly to the edge of the Leaf, is it even or ſcallop'd. Where theſe Inoſculations are not made, there we have no *Leaves*, but only a company of *Ramulets*, as in *Fennel*.

The Formations and Fouldings of Leaves have one Date, or are the contemporary works of Nature; each Leaf obtaining its diſtinct ſhape, and proper poſture together; both being perfect, not only in the outer, but Central and minuteſt Leaves, which ſometimes are five hundred times ſmaller than the outer; both which in the Cau-

tious

tious opening of a *Gormen* may be ſeen.

Nor is there greater Art in the Forms, than in the Foulds or Poſtures of *Leaves*; both anſwerably varying, as this or that way they may be moſt agreeable. Of the *Quincuncial* poſture, ſo amply inſtanc'd in by the Learned Dr. *Brown*, I ſhall omit to ſpeak. Others there are, which though not all ſo univerſal, yet equally neceſſary where they are; giving two general advantages to the Leaves, Elegancy and Security, *ſc.* in taking up, ſo far as their Forms will bear, the leaſt room; and in being ſo conveniently couch'd, as to be capable of receiving protection from other parts, or of giving it to one another; as for inſtance,

Firſt, There is the *Plain-Lap*, where the Leaves are all laid ſomewhat convexly one over another, but not plaited; being to the length, breadth and number of Leaves moſt agreeable; as in the Buds of *Pear-tree*, *Plum-tree*, &c. But where the Leaves are not thick ſet, as to ſtand in the *Plain-lap*, there we have the *Plicature*; as in *Roſe-tree*, *Strawberry*, *Cinquefoyl*, *Burnet*, &c. For the Leaves being here plaited, and ſo lying in half their breadth, and divers of them thus alſo collaterally ſet together, the thickneſs of them all, and half their breadth, are much alike dimenſions; by which they ſtand more ſecure within themſelves, and in better conſort with other *Germen-Growths* in the ſame Truſs. If the Leaves be much indented or jagg'd, now we have the ſame Duplicature; where there are divers Plaits in the ſame Leaf, or Labels of a Leaf, but in diſtinct Sets, a leſſer under a greater; as in *Tanſey*, &c. When the Leaves ſtand not collaterally, but ſingle, and

and that they are moreover very broad; then we have the *Multiplicature*; as in *Gooſeberries*, *Mallows*, &c. the Plaits being not only divers in the ſame Leaf, but of the ſame ſet continuant, and ſo each Leaf gather'd up in five, ſeven, or more Foulds, in the ſame manner as our Gentlewomens Fans: Where either the thickneſs of the Leaf will not permit a *flat lap*, or the fewneſs of their number, or the ſmallneſs of their Fibres, will allow the *Rowl*, there this may be obſerved; which is ſometimes ſingle, as in *Bears-Ears*; ſometimes double, the two *Rowls* beginning at each edge of the Leaf, and meeting in the middle. Which again, is either the *Fore-Rowl*, or the *Back-Rowl*. If the Leaf be deſign'd to grow long, now we have the *Back-Rowl*, as in *Docks*, *Primroſes*, &c. For the main Fibres, and that with a conſiderable part of the *Cortical Body* ſtanding prominent from the *Back-plain* of the Leaf, they thus ſtand ſecurely couch'd up betwixt the two *Rowls*; on whoſe ſecurity the growth of the Leaf in length depends. But *Bears-Ears*, *Violets*, &c. upon contrary reſpects, are rowled up inwards. Laſtly, there is the *Tre-Rowl*, as in *Fern*; the *Labels* whereof, though all rowled up to the *main Stem*, yet could not ſtand ſo firm and ſecure from the Injuries either of the Ground or Weather, unleſs to the *Rowls* in breadth, that by the length were ſuper-induc'd; the *Stalk* or *main Stem* giving the ſame protection here, which in other Plants by the Leaves, or ſome particular *Mantling*, is contriv'd.

For according to the Form and Foulding of every Leaf or *Germen*, is its protection order'd; about ſix wayes whereof may be obſerv'd; *ſc.* by *Leaves*, *Surfoyles*, *Interfoyles*,

Stalks,

Stalks, *Hoods* and *Mantlings*. To add to what we have above given, one or two Instances. Every Bud, besides its proper Leaves, is covered with divers Leafy *Pannicles* or *Surfoyls*; which, what the Leaves are to one another, are that to them all: For not opening except gradually, they admit not the Weather, Wet, Sun or Ayr, to approach the Leaves, except by degrees respondent, and as they are leisurely inur'd to bear them. Sometimes, besides *Surfoyls*, there are also many *Interfoyls* set betwixt the Leaves, from the Circumference to the Center of the *Bud*; as in the *Hasel*: For the Fibres of these Leaves standing out so far from a plain surface; they would, if not thus shelter'd, lie too much expos'd and naked to the Severities of the Weather. Where none of all the Protections above-named, are convenient, there the Membranes of the Leaves by continuation in their ſirſt forming (together with ſome Fibres of the *Lignous Body*) are drawn out into ſo many *Mantles* or *Veins*; as in *Docks*, *Snakeweed*, &c. For the Leaves here being but few, yet each Leaf and its Stalk being both exceeding long, at the bottom whereof the next following Leaf ſtill ſprings up; the form and poſture of all ſuch, as ſuperſedes all the other kinds of protection, and ſo each Leaf apart is provided with a Veil to it ſelf.

The Uſes of the Leaves, I mean in reſpect of their ſervice to the Plant it ſelf, are theſe; firſt, for Protection, which, beſides what they give to one another, they afford alſo to the *Flower* and *Fruit*: To the *Flower* in their Foulds; that being, for the moſt part, born and uſher'd into the open Ayr by the *Leaves*. To the *Fruit*, when afterwards

afterwards they are diſplay'd, as in *Strawberries*, *Grapes*, *Raſps*, *Mulberries*, &c. On which, and the like, ſhould the Sun-Beams immediately ſtrike, eſpecially while they are young, they would quite ſhrivel them up; but being by the Leaves ſcreened off, they impreſs the circumjacent Ayr ſo far only as gently to warm the ſaid Fruits, and ſo to promote their Fermentation and Growth. And accordingly we ſee, that the Leavs above-named are exceeding large in proportion to the *Fruits*: whereas in *Pear-trees*, *Apple-trees*, &c. the *Fruit* being of a ſolider *Parenchyma*, and ſo not needing the like protection, are uſually equal with, and often wider in Diameter than the *Leaves*.

Another uſe is for Augmentation; or, the capacity for the due ſpreading and ampliation of a Tree or Plant, are its Leaves: For herein the *Lignous Body* being divided into ſmall Fibres, and theſe running all along their lax and ſpongie *Parenchyma*; they are thus a Body fit for the imbibition of *Sap* and eaſie growth. Now the *Sap* having a free reception into the Leaves, it ſtill gives way to the next ſucceeding in the *Branches* and *Trunk*, and the voyding of the *Sap* in theſe, for the mounting of that in the *Root*, and ingreſs of that in the Mould. But were there no Leaves to make a free reception of *Sap*, it muſt needs be ſtagnant in all the parts to the *Root*, and ſo the *Root* being clogg'd, its fermenting and other Offices will be voyded, and ſo the due growth of the whole. As in the motion of a Watch, although the original term thereof be the Spring, yet the capacity for its continuance in a due meaſure throughout all the Wheels, is the free and eaſie motion of the Ballance.

Laſtly,

Laſtly, As the Leaves ſubſerve the more copious advancement, ſo the higher purity of the *Sap*: For this being well fermented both in the Root, and in its Aſcent through the *Trunk*, and ſo its Parts prepar'd to a farther ſeparation; the groſſer ones are ſtill depoſited into the Leaves; the more elaborate and eſſential only thus ſupplied to the *Flower*, *Fruit* and *Seed*, as their convenient Aliment. Whence it is, that where the *Flowers* are many and large, into which the more odorous Particles are copiouſly receiv'd, the green Leaves have little or no ſmell; as thoſe of *Roſe-tree*, *Carnations*, *French-Marigold*, *Wood-bind*, *Tulips*; &c. But on the contrary, where the Flowers are none or ſmall, the green Leaves themſelves are likewiſe of a ſtrong ſavour; as thoſe of *Wormwood*, *Tanſie*, *Baum*, *Mint*, *Rue*, *Geranium Moſchatum*, *Angelica*, and others.

An

An Appendix.

Of Thorns, Hairs and Globulets.

THorns are of two kinds, *Lignous* and *Cortical*. Of the first are such as those of the *Hawthorn*, and are constituted of all the same substantial Parts whereof the *Germen* it self, and in a like proportion: which also in their Infancy are set with the resemblances of divers minute Leaves. In affinity with these are the *Spinets* or *Thorny Prickles* upon the Verges and Tops of divers Leaves, as of *Barberry*, *Holly*, *Thistle*, *Furze*, and others; all which I think are the

the filamentous extremities of the *Lignous Body* ſheathed in the *Skin.*

Cortical Thorns are ſuch as thoſe of the *Rasberry* Buſh, being not, unleſs in a moſt extraordinary ſmall proportion propagated from the *Lignous Body*, but almoſt wholly from the *Cortical* and *Skin*, or from the *Barque.*

The growth of this *Thorn* may farther argue what in the Second Chapter we ſuppoſed; *ſc.* That as the proper tendency of the *Lignous Body*, is to aſcend, ſo of the *Cortical* to deſcend. For as the *Lignous Thorn*, like other Parts upon the *Trunk*, in its growth aſcends; this being almoſt wholly *Cortical*, pointeth downward. The uſe of *Thorns* the very Ingenious Mr. *Sharrock* obſerved.

Upon the Leaves of divers Plants two Productions ſhew themſelves, *ſc. Hairs* and *Globulets.* Of *Hairs*, only one kind is taken notice

tice of, althoegh they are various. Ordinarily they are plain; which when fine and thick ſet, as on moſt *Hairy Buds*; or fine and long, as on thoſe of the *Vine*, we call them *Down*.

But ſometimes they are not plain, but branched out, from the bottom to the top, reciprocally on every ſide, in ſome reſemblance to to a *Stags-Horn*; as in *Mullen*. And ſometimes they are *Aſtral*, as upon *Lavender*, and ſome other Leaves, and eſpecially thoſe of *Wild Olive*; wherein every *Hair* riſing in one round entire Baſis a little way above the Surface of the Leaf, is then diſparted, Star-like, into ſeveral, four, five or ſix points, all ſtanding at right Angles with the ſaid perpendicular Baſis.

The Uſes of Hairs are for Diſtinction and Protection. That of Diſtinction is but ſecondary, the Leaves being grown to a conſiderable

rable ſize. That of Protection is the prime, for which they were originally form'd together with the Leaves themſelves, and whoſe ſervice they enjoy in their Infant-eſtate: For the *Hairs* being then in form of a *Down*, alwayes very thick ſet, thus give that protection to the Leaves, which their exceeding tenderneſs then requires; ſo that they ſeem to be veſted with a Coat of *Frieze*, or to be kept warm, like young and dainty Chickens, in Wooll.

Globulets are ſeen upon *Orach*, both Garden and wild; and yet more plainly on *Mercury* or *Bonus Henricus*. In theſe, growing almoſt upon the whole Plant, and being very large, they are by all taken notice of.

But ſtrict Obſervation diſcovers, that theſe *Globulets* are the natural and conſtant Off-ſpring of very many other Plants. Both theſe

Globulets

Globulets, and likewiſe the diverſity of *Hairs*, I find the Learned Mr. *Hook* hath already obſerved. They are of two kinds; *Tranſparent*, as upon the Leaves of *Hyſop*, *Mint*, *Baume*, and many more: *White*, as upon thoſe of *Germander*, *Sage*, and others. All which, though the naked Eye will diſcover, yet by the help of Glaſſes we may obſerve moſt diſtinctly. The uſe of theſe we ſuppoſe the ſame with thoſe of the *Flower*, whereof we ſhall ſpeak.

CHAP.

CHAP. V.

Of the Flower.

WE next proceed to the *Flower*. The general Parts whereof are moſt commonly three; *ſc.* the *Empalement*, the *Foliation*, and the *Attire*.

The *Empalement*, whether of one or more pieces, I call that which is the utmoſt part of the *Flower*, encompaſſing the other two. 'Tis compounded of the three general Parts, the *Skin*, the *Cortical* and *Lignous Bodies*; each *Empaler* (where there are divers) being as another little Leaf; as in thoſe of a *Quince-Flower*, as oft as they happen to be overgrown, is well ſeen.

As likewiſe in the *Primroſe*, with the green Flower, commonly ſo call'd, though by a miſtake; for that which ſeems to be the *Flower*, is only the more flouriſhing *Empalement*, the *Flower* it ſelf being white; but the continuation of all the three aforeſaid Parts into each *Empaler*, is diſcoverable, I think, no where better than in an *Artichoke*, which is a true *Flower*, and whoſe *Empalers* are of that amplitude, as fairly to ſhew them all: As alſo, that the Original of the *Skin* of each *Empaler* is not diſtinct from that of the reſt; but to be all one piece, laid in ſo many Plaits or Duplicatures as there are *Empalers*, from the outermoſt to the inner and moſt central ones.

The Deſign of the *Empalement*, is to be ſecurity and Bands to the other two Parts of the *Flower*: To be their ſecurity before its opening, by intercepting all extremi-

ties

ties of Weather: Afterwards to be their Bands, and firmly to contain all their Parts in their due and moſt decorous poſture; ſo that a *Flower* without its *Empalement*, would hang as uncouth and taudry as a *Lady* without her *Bodies*.

Hence we have the reaſon why it is various, and ſometimes wanting. Some *Flowers* have none, as *Tulips*; for having a fat and firm Leaf, and each Leaf likewiſe ſtanding on a broad and ſtrong Baſis, they are thus ſufficient to themſelves. *Carnations*, on the contrary, have not only an *Empalement*, but that (for more firmitude) of one piece: For otherwiſe, the foot of each Leaf being very long and ſlender, moſt of them would be apt to break out of compaſs; yet is the top of the *Impalement* indented alſo; that the Indentments, by being lapp'd over the Leaves before their expanſion, may then protect them; and by being ſpred under them afterwards, may better ſhoulder and prop them up. And if the feet of the Leaves be both long and very tender too, here the *Empalement* is numerous, though conſiſting of ſeveral pieces; yet thoſe in divers Rounds, and all with a counterchangeable reſpect to each other (which alſo the Learned Dr. *Brown* obſerves) as in all *Knapweeds*, and other *Flowers*; whereby, how commodious they are for both the aforeſaid ends, may eaſily be conceiv'd; and well enough exemplified by the Scales of Fiſhes, whereunto, as to their poſition, they have not an unapt reſemblance.

The *Foliation* alſo, is of the ſame ſubſtantial nature with the green Leaf; the *Membrane*, *Pulp*, and *Fibres* whereof, being, as there, ſo here, but the continuation of the *Skin*, the *Cortical* and *Lignous Bodies*. The

The Foulds of the *Flower* or *Foliation* are various, as thoſe of the green Leaf; but ſome of them different. The moſt general are, Firſt, The *Plain Couch*, as in *Roſes*, and many other double *Flowers*. then the *Concave Couch*, as in *Blattaria flore albo*. Next the *Plait*, as in ſome of the Leaves of *Peaſe-Blooms*, in the Flowers of *Coriander*, &c. which is either ſingle, as in thoſe nam'd; or double, as in *Blew-Bottle*, *Jacea*, and more of that rank. Next, the *Couch* and *Plait* together in the ſame Flower, as in *Marigolds*, *Daiſies*, and all others of an agreeing form: where the firſt apparent Fould or Compoſture of the Leaves is in *Couch*; but the Leaves being erect, each likewiſe may be ſeen to lie in a double *Plait* within it ſelf. Then the *Rowl*, as in the *Flowers* of *Ladies-Bower*, the broad top of each Leaf being by a double *Rowl* foulded up in-

 wardly.

wardly. Next, the *Spire*, which it the beginning of a *Rowl*; and may be ſeen in the Flowers of *Mallows*, and others. Laſtly, the *Plait* and *Spire* together, where the part analogous to the *Foliation*, is of one piece, the *Plaits* being here laid, and ſo carried on by Spiral Lines to the top of the *Flower*, as is in divers, and I think in *Convotoulus Doronici folio* more elegantly ſeen. The reaſon of all which varieties, a comparative conſideration of the ſeveral parts of the *Flower* may ſuggeſt. Ile only mention, that no *Flower* that I find, hath a *Back-Rowl*, as hath the green Leaf, for two Reaſons; becauſe its Leaves have not their Fibres ſtanding out much on their backſide, as the green Leaves have; and becauſe of its Attire, which it ever emboſomes, and cannot ſo well do it by a *Back-Rowl*.

The uſual Protections of *Flowers*

ers by the Precedents are expreſs'd, *ſc. Green Leaves* and *Empalements*. Some have another more peculiar, that is a *double Vail*; as the *Spring-Crocus*. For having no *Empalement*, and ſtarting up early out of the Mould, even before its *Green Leaves*, and that upon the firſt opening of the Spring; leſt it ſhould thus be quite ſtarved, 'tis born ſwath'd up in a double Blanket, or with a pair of Sheets upon its Back.

The Leaves of divers *Flowers* at their Baſis have an *hairy Tuft*; by which *Tufts* the Concave of the *Empalement* is fill'd up; that, being very choice and tender, they may thus be kept in a gentle and conſtant warmth, as moſt convenient for them.

The Leaves of the *Flower*, though they are not hairy all over, yet in ſome particular parts they are often ſet with a fine Downy Velvet;

that, being by their ſhape and po-ſture in thoſe parts contiguous to their delicate and tender Attire, they may thus give it a more ſoftly and warmer touch. Thus in the Flower of *Ladies Bower*, thoſe parts of its Leaves which rowl in-ward, and lie contiguous to the Attire, are Downy; whereas the other parts are plain and ſmooth: So the Flowers of *Peaſe*, *Spaniſh Broom*, *Toad-Flax*, and many o-thers, where contiguous to their Attires, are deck'd with the like *Hairy Velvet*.

As upon the Green Leaves, ſo upon the Flowers are *Globulets* ſometimes ſeen; as upon the back-ſide of that of *Enula*. On none more plainly than that kind of *Blattaria* with the white Flower; where they are all tranſparent, and growing both on the Stalk and Leaves of the *Flower*, each ſhewing likewiſe its *Peduncle* whereon it is erected.

The

The uſe of the *Flower*, or the *Foliation* whereof we now ſpeak, (that is, as to its private ſervice) is for the protection of the Attire; this, as its under, and the *Empalement* as its upper Garments; as likewiſe of the *Fruit*: The neceſſity of which Service, in ſome Caſes, by the different ſituation of the *Flower* and *Fruit*, with reſpect to each other, is evident; *Apples*, *Pears*, and ſeveral other *Fruits*, ſtanding behind or under the *Flower*; but *Cherries*, *Apricots*, and divers others, within it; for theſe, being of a very tender and pulpous Body, and withal putting forth with the colder part of the Spring, could not weather it out againſt the Variations and Extremities of the Air, (as thoſe of a more ſolid *Parenchyma* can) except lodged up within their *Flowers*.

And as the *Flower* is ſerviceable to the ſafety of the *Fruit*, ſo is it to its growth; *ſc.* in its Infancy, or *Embryo*-eſtate ; for which purpoſe, as there is a Flower, ſo that Flower is greater or leſs, according as the nature of the Fruit to which it belongs, and the plenty of the *Sap* by which the Fruit is fed, doth require. Thus, where the young Fruit is of a ſolider conſtitution, and the aſcent of the *Sap* leſs copious, were there here no *Flower* to promote the ſaid aſcent thereof into the Fruit (in the manner as is effected by the Green Leaves) it muſt needs pine and die, or prove leſs kindly. On the contrary, ſhould the Flower be over-large, it would not only promote the aſcent of the *Sap* up to the Fruit, but being as yet over-proportionate to it, would likewiſe it ſelf exhauſt the ſame *Sap*, as faſt as aſcendent ; like a greedy Nurſe, that prepares the Meat for her Child, and then eats it up her ſelf. Thus we ſee

Apples and *Pears* with a *Flower* of a moderate ſize, like their Body; of a middle Conſtitution, and their *Sap* of a middle quantity: But *Quinces*, being more ſolid, beſides that they have as great a *Flower*, the *Impalers* of their *Flower* alſo thrive ſo far as to become handſom Leaves, continuing alſo after the *Flower* is fallen, firm and verdent a great while; ſo long till the *fruit* be able to provide for it ſelf. On the other hand, *Plums* being more tender and Sappy than *Apples* and *Pears*, beſides that their *Empalers* are much alike, their *flower* is leſs. and *Gooſeberries* and *Currans*, which are ſtill more Pulpy, and the courſe of the *Sap* towards them more free, have yet a *flower* far leſs. And *Grapes*, whoſe *Sap* is ſtill of quicker Aſcent, have ſcarce any *flower* at all; only ſome ſmall reſemblance thereof, ſerving juſt upon the ſetting of the *fruit*, and no longer.

The

The *Attire* I find to be of two kinds, *Seminie* and *Florie*: That which I call *Seminie*, is made up of two general parts, *Chives* and *Semets*, one upon each *Chive*. These *Semets* have the appearance (especially in many *flowers*) of so many little *Seeds*; but are quite another kind of Body: For upon enquiry we find, that these *Semets*, though they seem to be solid, and for some time after their first formation, are entire; yet are they really hollow; and their side, or sides, which were at first entire, at length crack asunder: And that moreover the Concave of each *Semet* is not a meer vacuity, but fill'd up with a number of minute Particles, in form of a Powder; which, though common to all *Semets*, yet in some, and particularly those of a *Tulip*, being larger, is more distinctly observable.

These *Semets* are sometimes fast-

ned

ned ſo, as to ſtand erect above their *Chive*, as thoſe of *Larks-heel*. Sometimes, and I think uſually, ſo as to hang a little down, in the manner and figure of a *Kidney*; as in *Mallows*. Their Cleft or Crack is ſometimes ſingle, but for the moſt part double: At theſe Clefts it is that they disburſe their Powders; which as they ſtart out, and ſtand betwixt the two Lips of each Cleft, have ſome reſemblance to the common Sculpture of a *Pomegranate* with its Seeds looking out at the Clefts of its *Rind*: This muſt be obſerv'd when the Clefts are recently made, which uſually is before the expanſion of the *Flower*.

The Particles of theſe Powders, though like thoſe of Meal or other Duſt, they appear not eaſily to have any regular ſhape; yet upon ſtrict obſervation, eſpecially with the aſſiſtance of an indifferent

Glaſs.

Glaſs, it doth appear, that they are nothing elſe but a *Congeries* of ſo many perfect *Globes* or *Globulets*: That which obſcures them; is their being ſo ſmall. In *Dogs-Mercury*, *Borage*, and very many more Plants, they are extreamly ſo. In *Mallows*, and ſome others, more fairly viſible.

Some of theſe Powders are yellow, as in *Dogs-Mercury*, *Goats-Rue*, &c. and ſome of other colours: But moſt of them I think are white; and thoſe of yellow *Henbane* very elegant; the disburs'd Powders whereof, to the naked eye, are white as Snow; but each *Globulet*, through a Glaſs, tranſparent as Cryſtal; which is not a fallacy from the Glaſs, but what we ſee in all tranſparent Bodies whatſoever, lying in a Powder or ſmall Particles together.

The *Florid Attire*, is commonly

ly known by the blind and rude Name of *Thrums*; as in the Flowers of *Marigold*, *Tansie*, &c. How adequate its imposition is, observation will determine: For the several *Thrums* or rather *Suits*, whereof the *Attire* is made up, however else they may differ in various Flowers, in this agree, that they are never consistent of more than one, sometimes of two, and for the most part of three pieces (for which I call them *Suits*) and each piece of a different, but agreeable and comely form.

The *outer part* of every *Suit*, is its *Floret*: whose *Body* or *Tube* is divided at the top (like that of the *Cowslip*) into divers distinct Leaves; so that a *Floret* is the Epitome of a *flower*; and is all the *flower* that many Plants, as *Mugwort*, *Tansie*, and others, have. What the Learned Dr. *Brown* observeth of the number Five as to

the

the Leaves of the *flower*, is ſtill more univerſally holding in theſe of the *Floret*.

Upon the Expanſion of the *Floret*, the next part of the *Suit* is from within its *Tube* brought to ſight; which we may (with reſpect to that within it) call the *Sheath*: For this alſo, like the *Floret*, is a concave Body; in its ſhape very well reſembling the Fiſtulous Pouches of *Wake-Robin*, or of *Dragon*.

The *Sheath*, after ſome time, dividing at the top, from within its Concave, the third and innermoſt part of the *Suit*, *ſc.* the *Blade* advanceth and diſplayes it ſelf. This part is not hollow, as the other two, but ſolid; yet at its point, not originally, but after ſome time, is evermore divided into two halves.

Upon the diviſion of the ſaid Point, there appears, as upon the opening

opening of a *Semet*, a Powder of *Globulets*, which before lay enclosed up within its Clefts; and are of the ſame nature with thoſe of a *Semet*, though not ſo copious: So that all *flowers* have their *Powders* or *Globulets*. The whole *Attire* may in *Knapweed*, *Blewbottle*, &c. be obſerved.

The uſe of the *Attire*, how contemptibly ſoever we may look upon it, is certainly great. And though for our own uſe we value the Leaves of the *Flower*, not the *Foliation*, moſt; yet of all the three Parts, this in ſome reſpects is the choyceſt, as for whoſe ſake and ſervice the other two are made. The uſe hereof, as to Ornament and Diſtinction, is unqueſtionable; but is not all. As for Diſtinction, though by the help of Glaſſes we may make it to extend far; yet in a paſſant view, which is all we uſually make, we cannot ſo well. As for Ornament, and particularly in reference to the *Semets*, we may ask, If for that meerly theſe were meant, then why ſhould they be ſo made as to break open, or to contain any thing within them? Since their Beauty would be as good as if they were not hollow, and is better before they crack and burſt open, than afterwards.

A farther uſe hereof therefore we muſt acknowledge, and may obſerve; and that is for food; for Ornament and Diſtinction to us, and for Food to other Animals. I will not ſay, but that it may ſerve even to theſe for Diſtinction too, that they may be able to know one Plant from another, and in their flight or progreſs ſettle where they like beſt; and that therefore the varieties of theſe ſmall parts are many, and well obſerved by them, which we take no notice of: Yet the

the finding out of Food is but in order to enjoy it: Which, that it is provided for a vaſt number of little Animals in the *attires* of all *Flowers*, obſervation perſwades us to believe. For why elſe are they evermore here found? Go from one Flower to another, great and ſmall, you ſhall meet with none untaken up with theſe Gueſts. In ſome, and particularly the *Sun-flower*, where the parts of the *Attire*, and the *animals* for which they provide, are larger, the matter is more viſible. We muſt not think, that God Almighty hath left any of the whole Family of his Creatures unprovided for; but as the Great Maſter, ſome where or other carveth out to all; and that for a great number of theſe little Folk, He hath ſtored up their peculiar proviſions in the *Attires* of *Flowers*; each *Flower* thus becoming their Lodging and their

Dining-Room, both in one.

Wherein the particular parts of the *Attire* may be more diſtinctly ſerviceable, this to one Animal, and that to another, I cannot ſay: Or to the ſame Animal, as a *Bee*, whether this for the *Honey*, another for their *Bread*, a third for the *Wax*: Or whether all only ſuck from hence ſome *Juice*; or ſome may not alſo carry ſome of the Parts, as of the *Globulets*, wholly away: Or laſtly, what may be the primary and private uſe of the *attire* (for even this aboveſaid, though great, yet is but ſecondary) I now determine not.

CHAP.

CHAP. VI.

Of the Fruit.

THe general compoſition of all *Fruits* is one, that is, their *Eſſential* and truly *Vital Parts*, are in all the ſame, and but the continuation of thoſe which in the other Parts of a *Vegetable*, we have already obſerved: Yet becauſe by the different Conſtitutions and Tinctures of theſe Parts, divers conſiderably different Fruits reſult; I ſhall therefore take a particular view of the more known and principal of them, *ſc. Apples*, *Pears*, *Plums*, *Nuts* and *Berries*.

An *Apple*, if cut traverſe, appears conſtituted of four diſtinct

Parts, the *Pill*, the *Parenchyma*, *Branchery* and *Coare*. The *Pill* is only the ſpreading and dilatation of the skin, or utmoſt part of the Barque in the Branch. The *Parenchyma*, when full ripe, is a tender delicate Meat: Yet as the *Pill* is but the continuation of the utmoſt part of the *Barque*; ſo is this but the continuance and ampliation, or (as I may call it) the ſwelth and ſuperbience of the *Inner part* thereof; which upon obſervation of a young and Infant-*Apple* eſpecially, is evident. Thus we ſee the *Pith*, which is often tough, in many Roots, as *Parſneps*, *Turneps*, &c. is tender and edible. So here, the *Parenchyma*, though originally no more than the *Barque*, yet the plenty and purity of its *Sap* being likewiſe effectual to the fulneſs and fineneſs of its growth, it thus becomes a ſoft and tender meat. The *Branchery* is no-

nothing elſe but the Ramifications of the *Lignous Body* throughout all the parts of the *Parenchyma*; the greater Branches being likewiſe by the *Inoſculations* of the leſs (as in the Leaf) united together. The main Branches are uſually fifteen; ten are ſpred and diſtributed through the *Parenchyma*, all enarching themſelves towards the *Cork* or *Stool* of the *Flower*; the other five running from the *Stalk* in a directer Line, at laſt meet the former at the ſaid *Cork*, and are there oſculated with them. Theſe five are originated from one; which running along the Center of the *Stalk*, and part of the *Parenchyma* of the *Fruit*, is therein at laſt divided. To theſe the Coats of the *Kernels* are faſtned; ſo that whereas theſe Branches were originally all extended even beyond the *Fruit*, and inſerted into the *Flower* for the due growth thereof;

the *Fruit* afterwards growing to ſome head, and ſo intercepting and preying upon the Aliment of the *Flower*, ſtarves that, and therefrom ſuperſedes the ſervice of the ſaid Branches to it ſelf, ten for its *Parenchyma*, and five for its *Seed*. The *Coar* is originated from the *Pith*; for the *Sap* finding room enough in the *Parenchyma*, through which to diſpence it ſelf all abroad, quits the *Pith*, which thereby hardens into a *Coar*. Thus we ſee the *Inſertions*, although originate from the *Cortical Body*, yet their Parts being, by the *Inoſculations* of the *Lignous*, ſo much compreſs'd and made to co-incide together, they become a Body very compact and denſe. And in the *Barque* we ſee the ſame effect by *arefaction* only, or a meer *voydance* of the *Sap*; the *Inner Part* whereof, though ſoft and ſappy, yet its ſuperficial *Rind* is often ſo hard and ſmooth, that

that it may be fairly writ upon.

In a *Pear* there are five distinct Parts, the *Pill*, the *Parenchyma*, *Branchery*, *Calculary* and *Acetary*. The three former are here and in an *Apple* much alike; saving that here the *Inner* or *Seed-Branches* are ordinarily ten. The *Calculary* (most observable in rough-tasted, or *Choak-Pears*) is a *congeries* of little stony Knots: They are many of them dispersed throughout the whole *Parenchyma*; but lying more continuous and compact together towards the Center of the *Pear*, surround the *Acetary* there in a somewhat Globular Form. About the *Stalk* they stand more distant; but towards the *Cork* or *Stool* of the *Flower*, they still grow closer, and there at last gather (almost) into the firmitude of a *Plum-stone* it self. Within this lies the *Acetary*: 'tis of a soure tast, and by the bounding of the *Calculary* of of a *Globular Figure*. 'Tis a ſimple Body, having neither any of the *Lignous* branched in it, nor any Knots. It is of the ſame ſubſtantial nature with the *Parenchyma*; but whether it be abſolutely one with it, or be derived immediately from the *Pith*, my Enquiries yet made, determine not.

The Original of the *Calculary* I ſeem to have neglected: But hereof we may here beſt ſay, that whereas all the other Parts are Eſſential and truly *Vital*, the *Calculary* is not; but that the ſeveral Knots whereof it conſiſts, are only ſo many meer Concretions or Precipitations out of the *Sap*; as in *Urines*, *Wines*, and other *Liquors*, we often ſee. And that this *Precipitation* is made by the mixture and re-action of the Tinctures of the *Lignous* and *Cortical Bodies* upon each other: Even as all *Vegetable Nutrition* or *Fixation* of of Parts is alſo made by the joynt efficiency of the two ſame Tinctures, as hath been ſaid. Hence we find, that as the *Acetary* hath no Branches of the *Ligneus Body*, ſo neither hath it any Knots. Hence likewiſe it is, that we have ſo different and contrary a taſte in the *Parenchyma* beyond the *Calculary*, from that in the *Acetary*; for whereas this is ſoure, that, wherein the ſaid *Precipitations* are made, is ſweet; being much alike effect, to what we find in mixing of *Corals*, &c. with *Vinegar* or other *acid Liquor*.

In a *Plum* (to which the *Cherry*, *Apricot*, *Peach*, *Walnut*, &c. ought to be referr'd) there are four diſtinct Parts, the *Pill*, the *Parenchyma*, *Branchery* and *Stone*. The *Pill* and *Parenchyma* are, as to their Original, with thoſe of an *Apple* or *Pear* both alike: As likewiſe the *Branchery*; but differently

rently ramified. In *Plums* (I ſuppoſe all) there are five main *Out-Branches*, which run along the Surface of the *Stone* from the *Baſis* to the point thereof, four of them by the one Ridge. and one by the other oppoſite to it. In an *Apricot* there is the ſame number, but the ſingle Branch runs not upon the Surface, bnt through the Body of the *Stone*. There are likewiſe two or three ſmaller Branches, which run in like manner under the other Ridge for ſome ſpace, and then advancing into the *Parenchyma*, therein diſperſe themſelves: Theſe latter ſort in *Peaches* are numerous throughout: But notwithſtanding the different diſpoſition of the Branches of the *Fruits* aforeſaid; yet is there one Branch diſpos'd in one and the ſame manner in them all: The entrance hereof into the *Stone* is at its *Baſis*; from whence running through

its

its Body, and ſtill inclining or arching it ſelf towards its Concave, is at laſt about its Cone thereinto emergent, where the Coats of the *Seed* are appendent to it. Of the *Seed-Branch* 'tis therefore obſervable that after its entrance into the *Fruit*, 'tis alwaies prolonged therein to a conſiderable length; as is ſeen not only in *Apples*, &c. where the *Seed* ſtands a good diſtance from the *Stalk*; but in *Plums* likewiſe, where it ſtands very near it; in that here the *Seed-Branch*, as is ſaid, never ſtrikes through the *Stone* into the Coats of the *Seed* directly, but about its Cone or remoter end. The *Stone*, though it ſeem a ſimple Body, yet it is compounded of different ones: The Inner Part thereof, as it is by far the thinneſt, ſo is it the moſt denſe, white, ſmooth and ſimple. The Original is from the *Pith*; difficult, but curious to obſerve: For the

Seed-

Seed-Branch, not ſtriking directly and immediately quite through the *Baſis* of the *Stone*, but in the manner as is above deſcribed, carries a conſiderable part of the *Pith*, now gather'd round about it, as its *Parenchyma*, along with it ſelf; which, upon its entrance into the concave of the *Stone* about its farther end, is there in part ſpred all over it, as the Lining thereof. The outer and very much thicker Part conſiſteth partly of the like *Precipitations* or concrete Particles, as in a *Pear*, being gathered here much more cloſely, not only to a Contiguity, but a coalition into one entire Stone; as we ſee in *Pears* themſelves, eſpecially towards the *Cork*, they gather into the like Stonineſs; or as we ſee a *Stone*, *Mineral* or *Animal*, oftentimes the product of accumulated *Gravel*: But as the *Parenchyma* is mixed with the Concretion in the *Calcu-*

lary,

lary, ſo is it alſo, though not viſibly, with theſe in the *Stone*, the ground of the *Stone* being indeed a perfect *Parenchyma*; but by the ſaid Concretions ſo far alter'd, as to become dry, hard and undiſtinguiſhable from them.

In a *Nut* (by which an *Achorn* is analogous) there are three general Parts, the *Cap*, *Shell* and *Pith*. The *Cap* is conſtituted of a *Pill* and *Parenchyma* derived from the *Barque*, and *Ramulets* from the *Lignous Body* of the *Branch*. The *Shell* likewiſe is not one ſimple Body, but compounded. The Superficial Part thereof is originated from the *Pill* or *Skin* of the *Cap*, from the inſide whereof it is in a Duplicature produc'd and ſpred over the *ſhell*: which, if you look at the *Baſis* of the *ſhell*, is farther evident; for that being continuous with the *Parenchyma* of the *Cap*, without the interpoſure of of the *Skin*, the said superficial Part is there wanting. The thicker and inner part of the *shell* consisteth of the same *Parenchyma* as that of the *Cap*, with a *congeries* of *Precipitations* filled up, as in a Stone. And as the *Lignous Body* is branched in a Stone, so, with some difference, in a Shell. The *Outer Branches* or *Ramulets* are numerous, each issuing out of the *Parenchyma* of the *cap*, and entring the *Shell* at the Circumference of its *Basis*, and so running betwixt its superficial and inner parts towards its *cone*, in a Round, The *Inner* or *Seed-Branch* is single, entering in, as do the other, not at the *Basis* of the *Shell*, but at the *center* thereof; from whence it runs, not through the *Shell*, as in *Plums* through the *Stone*; but through the *Pith*, as far as the *cone*, where the Coats of the *Seed* hang appendent to it. The *Pith*, whether deri-

derived from the ſame part both in name and nature in the *Branch* and *Stalk*, or from the *Cortical Body*, I yet determine not.

A *Berry*, as a *Gooſeberry* (to which *Currans*, *Grapes*, *Hipps*, &c. are to be referr'd) conſiſteth, beſides the *Seed*, of the three general Parts, *Pill*, *Parenchyma* and *Branchery*: The *Pill* is originated as in the foregoing Fruits. The *Parenchyma* is double, as likewiſe in ſome other *Berries*: The *outer* is commonly, together with the *Pill*, call'd the *Skin*, and is that part we ſpit out, being of a ſoure taſte. As the *Pill* is originated from the *outer*, ſo this from the *inner part* of the *Barque*; and accordingly the Pores thereof may be obſerved plainly of a like ſhape with thoſe both of the *Cortical Body* and *Pith*. The *inner* is of a ſweet taſte, and is the part we eat: It is of a conſtitution ſo laxe and tender, as it would ſeem to be only a thicker or jellied Juice; although this likewiſe be a true *Parenchyma*, ſomething like that of an *Orange* or *Limon*, with its Pores all fill'd up with Liquor. The *Branchery* is likewiſe double: The *Exterior* runs betwixt the *Pill* and outer *Parenchyma* in arched Lines, from the *Stalk* to the *Stool* of the *Flower*. Theſe outer Branches, though of various number at the *Stalk*, yet at the *Cork* are uſually ten principal ones; five for the five Leaves of the *Flower*, and five for the *Chives*. The inner main Branches are two, diametrically oppoſite to each other, and at the *Cork* with the other inoſculated. From theſe two are branched other ſmaller, every one having a *Seed* appendent to it, whoſe Coats it entreth by a double Filament, one at the *Baſis*, the other at the *Cone*. They are all very white and turgent,

gent, and by a ſlaunt cut, may be obſerv'd concave; thus repreſenting themſelves analogous to ſo many true *ſpermatick Veſſels.*

The Uſes of *Fruits* are for *Man*, (ſometimes alſo other *Animals*, as are *Akerns* and *Haws*) and for the *Seed*. For Man, they are ſo variouſly deſirable, that till our Orchards and Store-Chambers, *Confectioners* Stores and *Apothecaries* Shops, our Ladies Cloſets, their Tables or Hands are empty of them, I ſhall not need to enquire for what. If it be asked, how the Fruit becomes, generally above all the other Parts, ſo pleaſant a Meat? It is partly from the *Sap*, the groſſer portion thereof being depoſited in the Leaves, and ſo the purer hereunto reſerved; partly from the Globular Figure of the Fruit; for the *Sap* being thus in a greater quantity herein, and in all parts equally diffus'd, the Concoction hereof is with greatest advantage favoured and promoted. Wherefore all Fruits which we eat raw, how small soever, are of a Globular form, or thereunto approaching; and the nearer, the delicater; amongst *apples*, the *Peppin*; amongst *Pears*, the *Burgundian*; and amongst all Fruits, the *Grape*; and amongst *Grapes*, the roundest, are of all the most dainty.

The visible cause of this Globular Figure, is the *Flower*; or the Inosculation of all the main Branches at the *Stool* of the *Flower*; and upon the fall of the *Flower*, the obtuseness, and with Wind and Sun, as it were the searing of their several ends: For thus the *Sap* entring the *Fruit*, being not able to effect, either a Disunion, or a shooting forth of the said Branches, and so to carry on their growth in length; they must thus of necessity be be enarch'd, and with the *Parenchyma* more and more expand themſelves. Whereas were they diſpos'd and qualified otherwiſe, than as is ſaid, inſtead of forming a Fruit within bounds, they would run out into all extravagance, and even into another little Tree or Leafy growth.

To the *Seed*, the *Fruit* is ſerviceable; Firſt, in order to its being ſupply'd with a due and moſt convenient *Sap*, the greater and leſs elaborated part thereof being, in its paſſage towards the *Seed*, thereinto received; the *Fruit* doing the ſame office to the *Seed*, which the *Leaves* do to the *Fruit*; the *Sap* in the *Fruit* being in a laxe compariſon, as the *Wine*; and that for the *Seed*, a ſmall part of the higheſt Spirit rectified from it.

So likewiſe for its Protection, in order to the proſperous carrying on and perfecting of its generation, and ſecurity being perfected. Which protection it gives not only to the Seminal *Sap* and *Seed* it ſelf, but alwaies alſo to its *Seed-Branch*. Thus we ſee an *Apple*, beſides that it is it ſelf of ample compaſs, for the ſake of its *Seed*, hath likewiſe its *coar*; as if it were not ſufficient, that the Walls of their Room are ſo very thick, unleſs alſo wainſcotted. In a *Pear* again, where the *Parenchyma* is of leſs compaſs than that of an *Apple*, to what protection this affords, that of the *Calculary* is ſuper-added. But in a *Plum*, where the *Parenchyma* is exceeding tender, and in a *Peach*, which hangs late, and till Autumn Froſts approach, we have not only the Rubbiſh of a *Calculary*, but ſtout Stone-Walls. Within which alſo, not only the *Seed* it ſelf, but the *Seed-Branch* is evermore immur'd. Laſtly, in a *Nut*, where the *ſhell* being not ſurround-

ed

ed with a *Parenchyma*, that protection is wanting without, 'tis answer'd by an ample *Pith* within it; and the *seed-Branch* likewise included, not meerly in the Body of the Shell, as in a *Plum*, but within the *pith* it self. So necessary is this design, that what the Hen by Incubation or Hovering, is to the Egg or Chick; that the whole *Fruit*, by comprehension, is to the *Seed*.

CHAP. VII.

Of the Seed.

AS the Original, ſo the Ultimate end & Perfection of *Vegetation* is the *Seed*. How it is the former, and in its ſtate apt for *Vegetation*, hath already been ſeen. How the latter, and in its ſtate of Generation, we ſhall now laſtly enquire. In doing which, what in the other ſtate was either not diſtinctly exiſtent, or not ſo apparent, or not ſo intelligible, will occur.

The two general Parts of the *Seed* are its *Covers* and *Body*. The *Covers* in this eſtate are uſually four;

four; the outmost we may call the *Case*: 'Tis of a very various form; sometimes a *Pouch*, as in *Nasturtium*, *Cochlearia*; a *Cod*, as in all *Pulse*, *Galega*; sometimes not entire, but parted, or otherwise open, as in *Sorrel*, *Knotgrass*, with many other forms; I think alwaies more heterogeneous to that of the *Seed*, by which it differs from the proper Coats. To this the Caps of *Nuts*, and the *Parenchyma's* of Fruits are analogous.

The two next are properly the Coats: In a *Bean* especially, and the like; from whence to avoyd Confusion, the denomination may run common to the responding Covers of other *Seeds*. The Colour of the outer is of all degrees, from White to the Blackness of *Jett*: Its Figure sometimes Kidney'd, as in *Alcéa*, *Behen*, *Poppy*; triangular, as in *Polygonatum*, *Sorrel*; triangular spherical, in *Mentha*,

tha, *Melissa*; circular, in *Leucoium*, *Amaranthus*; globular, in *Napus*, *Asperula*; oval, in *Speculum Veneris*, *Tithymalus*; half Globe, in *Coriander*; that which we take for one single round *Seed*, being a Conjugation of two; half Oval, in *Anise*, *Fennel*; Hastal, in *Lactuca*; Cylindrical, as, if I mistake not, in *Jacobæa*; Pyramidal, in *Geranium*, *Althææ Fol.* with many other differences: But the Perfection of one or two of the said Figures lieth in the *Case*: So that as all Lines and Proportions are in the *Flower*, so all Regular Figures in the *Seed*, or rather in its *Covers*.

'Tis sometimes glistering, as in *Speculum Veneris*; Rough-cast, in *Catanance*; Studded, in *Behen*, *Blattaria*; Favous, in *Papaver*, *Antirrhinum*, *Lepidium annuum*, *Alcea Vesicaria*, *Hyosciamus*, and many more, before the *Seeds* have lain

lain long by; Pounced, in *Phalangium Cretæ*, *Lithospermum*; Ramified, in *Pentaphyllum fragiferum*, *Erectum majus*, resembling the Fibres of the Ears of the Heart; some just *Quinquenerval*, as in *Anisum*, and many more, the *Lignous Body* being in five main Fibres branched therein.

The Covers of not only *Quince*-Seeds, and those of *Psyllium* (more usually taken notice of) but those also of *Horminum*, *Nasturtium*, *Eruca*, *Camelina*, *Ocymum*, and divers others, have a *Mucilage*; which, though it be not visible when the Seeds are throughly dry; yet lying a while in some warm Liquor, or only on the Tongue, it swells more or less, and upon them all fairly shews it self. On that of *Ocymum* it appears grayish; on the other, transparent; and on that of *Nasturtium Hortense* very large; even emulous of the inner Pulp surrounding

rounding a *Goosberry-seed.* The putting of *Clary-seed* into the Eye, may have been brought into use from this *Mucilage*, by which alone it may become Medicinal. And thus far of the *Superficies.*

The nature of the outer Coat is various, *Membranous*, *Cartilaginous* and *Stony*; the like *Precipitations* being sometimes made herein, as in a Stone or Shell; as in that of the Seeds of *Carthamum*, *Lithospermum*, and others. The Designment hereof, being either with respect to the *Seed* in its state of Generation; as where the Case is either wanting, or at least insufficient of it self, there for its due protection and warmth; or, in its state of *Vegetation*, for the better Fermenting of its Tinctures and *Sap*; the Fermentations of some *Seeds* not well proceeding, unless they lie in their Stony Casks in the Mould, like Bottled Liquors in Sand.

All

All *Seeds* have their outer Covers open; either by a particular *Foramen*, as in *Beans*, and other *Pulse*, as is ſaid; or by the breaking off of the *Seed* from its *Peduncle* or *Stool*, as in thoſe in *Cucumber*, *Cycory*; or by the entering and paſſage of a *Branch* or *Branches*, not only into the Concave thereof near the Cone, but alſo through the Cone it ſelf; as in *Shells* and *Stones*.

For the ſake of this *aperture* it is, that *Akerns*, *Nuts*, *Beans*, *Cucumbers*, and moſt other *Seeds*, are in their formation ſo placed, that the *Radicle* ſtill ſtandeth next to it; that, upon *Vegetation*, it may have a free and ready paſſage into the Mould.

The Original of the outer Coat, though from Parts of the ſame ſubſtantial nature, yet is differently made. In a *Plum*, the *Seed-Branch* which runs, as is deſcribed, through the

the Stone, is not naked, but, as is ſaid, inveſted with a thin *Parenchyma*, which it carries from the Stalk along with it; and which, by the *Ramification* of the ſaid *Branch* within the Stone, is in part dilated into a Coat. That of a *Bean* is from the *Parenchyma* of the Cod; the ſuperficial part of which *Parenchyma*, upon the large *peduncle* of the *Bean* becoming a thin Cuticle, and upon the *Bean* it ſelf a *cartilaginous* Coat.

The Original of the inner Coat of the *Bean* is likewiſe from the inner part of the ſaid *parenchyma*; which firſt is ſpred into a long Cake, or that which with the *ſeed-Branch* maketh the *peduncle* of the *Bean*; under which Cake, there is uſually a black part or ſpot; by the length of which, the inner part of the Cake is next inſerted into the outer Coat, and ſpred all over the Concave thereof.

This

This inner Coat, though when the *Seed* is grown old and dry, 'tis ſhrunk up, and in moſt Seeds ſo far as ſcarcely to be diſcern'd; yet in its firſt and juvenile Conſtitution, is a very Spongy and Sappy Body; and is then likewiſe (as the *Womb* in a pregnant Animal) in proportion very thick and bulky; in a *Bean*, even as one of the *Lobes* it ſelf: And in a *Plum* or *apricot*, I think I may ſafely ſay, half an hundred times thicker than afterwards, when it is dried and ſhrunk up; and can ſcarcely be diſtinguiſhed from the upper Coat. Upon which Accounts it is, in this eſtate, a true and fair *Parenchyma*.

In this Inner Coat in a *Bean*, the *Lignous Body* or *Seed-Branch* is diſtributed: Sometimes, as in *French-Beans*, throughout the whole Coat; as it is in a Leaf: In the Great *Garden-Bean*, upon its

its firſt entrance, it is bipartite, and ſo in ſmall Branches runs along the Circumference of the Coat, all meeting and making a kind of Reticulation againſt the Belly of the *Bean*. In the ſame manner the main Branches in the outer Coat of a *Kernel*, circling themſelves on both hands from the place of their firſt entrance, at laſt meet, and mutually inoſculate.

So that all the Parts of a *Vegetable*, the *Root*, *Trunk*, *Branch*, *Leaf*, *Flower*, *Fruit* and *Seed*, are ſtill made up of two ſubſtantially different Bodies.

And as every Part hath two, ſo the whole *Vegetable* taken together, is a compoſition of two only, and no more: All properly Woody Parts, Strings and Fibres, are one Body: All ſimple *Barques*, *Piths*, *Parenchyma's* and *Pulps*, and as to their ſubſtantial Nature, *Pills* and *Skins* likewiſe, all but one Body:

Body: the ſeveral Parts of a *Vegetable* all differing from each other, only by the various Proportions and Mixtures, and variouſly ſized Pores of theſe two Bodies.. What from theſe two general Obſervations might reaſonably be inferr'd, I ſhall not now mention.

The fourth and innermoſt Cover we may call the *Secondine*; the ſight whereof, by cutting off the Coats of an *Infant-Bean*, at the Cone thereof in very thin Slices, and with great Caution, may be obtain'd. While unbroken, 'tis tranſparent; being torn and taken off, it gathers up into the likeneſs of a Jelly, or that we call the *Tredle* of an Egg, when over-boyl'd. This *Membrane* in larger or elder *Beans*, is not to be found diſtinct; but becomes as it were the Lining of the inner Coat: But (as far as our Enquiries yet diſcover) it may in moſt other *Seeds*, even full grown, be

 diſtinctly

distinctly seen; as in those of *Cucumber*, *Colocynthis*, *Burdock*, *Carthamum*, *Gromwel*, *Endive*, *Mallows*, &c. 'Tis usually so very thin, as in the above-nam'd, as very difficultly to be discover'd. In some *Kernels*, as of *Apricots*, 'tis very thick; and in some other Seeds. That all these have the Analogy of one and the same Cover, which I call the *Secondine*, is most probably argu'd from their alike Natures; being all of them plain simple *Membranes*, with not the least Fibre of the *Lignous Body* or *Seed Branch*, visibly distributed in them; as also from their Contexture, which is in all of them more close.

The Concave of this *Membrane* is filled with a most transparent Liquor, out of which the Seed is formed; as in cutting a *petite* and *Infant-Bean*, may be seen; and yet better in a young *Walnut*. In

Beans

Beans I have obſerved it to turn, upon boyling, into a tender white *coagulum*.

Through this *Membrane*, the *Lignous Body* or *Seed-Branches* diſtributed in the inner Coat, at laſt ſhoot downright two ſlender Fibres, like two Navles, one into each *Lobe* of the *Bean*. The places where the ſaid Fibres ſhoot into the *Lobes*, are near the *Baſis* of the *Radicle*; and by their Blackiſhneſs well enough remark'd: but the Fibres themſelves are ſo very ſmall, as ſcarcely to be diſcern'd: Yet in a *Lupine*, of the larger kind, both the places where the Navel-Fibres ſhoot into the *Lobes* (which here from the *Baſis* of the *Radicle* is more remote) and the Fibres themſelves, are fairly viſible. For the *Seed-Branch*, upon its entrance into the Coat of the *Lupine*, is preſently divided into two *main Branches*, and thoſe two into other leſs;

whereof ſome underly, others aloſt, run along the Coat, and towards its other end meet and are inoſculated; where about two oppoſite, ſhallow, round, and moſt minute Cavities, anſwerable to two Specks of a *cartilaginous* gloſs, one in either *Lobe*, may be obſerved; which Specks are the ends of the ſaid *Navel-Fibres*, upon the ripening of the *Seed* there broken off. Theſe Fibres, from the Superficies of each *Lobe*, deſcend a little way directly down; preſently, each is divided into two Branches, one diſtributed into the *Lobes*, the other into the *Radicle* & *Plume*, in the manner as in the firſt Chapter is deſcrib'd. And thus far the Hiſtory. I ſhall now only with a brief account of the *Generation* of the *Seed*, as hereupon dependent, conclude this Diſcourſe.

Let us ſay then; that the *Sap* having in the *Root*, *Trunk* and *Leaves*, paſſed divers Concoctions and Separations, in the manner as they are ſaid to be perform'd therein; 'tis now at laſt, in ſome good maturity, advanced towards the *Seed*.

The more copious and cruder part hereof is again ſeperated by a free reception into the *Fruit*, or other Part analogous to it: being either ſufficiently ample to contain it, or at leaſt laxe enough for its tranſpiration, and ſo its due diſcharge. The more Eſſential part is into the *Seed-Branch* or Branches entertain'd; which, becauſe they are evermore of a very conſiderable length, and of a Conſtitution very fine, the ſaid *Sap* thus becomes in its Current therein, as in the *Spermatick Veſſels*, ſtill more mature.

In this matúre eſtate, from the *ſeed-Branch* into the Coats of the *ſeed*, as into the Womb, 'tis next delivered up. The meaner Part hereof again, to the outer, as *aliment* good enough, is ſupplied. The finer part is tranſmitted to the Inner; which, being, as is ſaid, a *Parenchymous* and more ſpacious Body, the *Sap* therefore is not herein, as in the outer, a meer *aliment*; but in order to its being, by Fermentation, farther prepared.

Yet the outer Coat, being on the contrary hard and denſe; for that reaſon, as it admitteth not the Fermentation of the *Sap* ſo well within it ſelf; ſo doth it the more promote and favour it in the Inner, being Bounds both to it and its *Sap*; and alſo quickneth the proceſs of the whole Work in the formation of the *Seed*.

Nor

Nor doth the outer Coat, for the ſame reaſon, more promote than declare the purity of the *Sap* now contained in the Inner: For being more hard and denſe, and ſo not perſpirable, muſt needs ſuppoſe the Parts of the *Sap* encompaſſed by it, ſince thus uncapable of any evacuation, to be therefore all, ſo choice, as not to need it.

The *Sap* being thus prepared in the inner Coat, as a Liquor now apt to be the *Subſtratum* of the future *Seed-Embrio*, by freſh ſupplies, is thence diſcharg'd; yet that it may not be over-copious; which, becauſe of the laxity of the Inner Coat from whence it iſſues, it might eaſily be; therefore as the ſaid inner Coat is bounded without by the upper Coat, ſo by the *Secondine* or *Membrane* is it bounded within; through which *Membrane* the *Sap* being filtr'd, or, as it were, tranſpiring, the depoſiture hereof, anſwerable to the *Colliquamentum* in an Egg, or to the *ſemen Mulibre*, into its Concave at laſt is made.

The other Part of the pureſt *ſap* emboſom'd in the Ramulets of the *ſeed-Branch*, runs a Circle, or ſome progreſs therein; and ſo becomes, as the *Semen Maſculinum*, yet more elaborate.

Wherein alſo, leſt its Current ſhould be too copious or precipitate, by their co-arcture and divarication where they are inoſculated, it is retarded; the nobleſt portion only obtaining a paſs.

With this pureſt *ſap*, the ſaid *Ramulets* being ſupplied, from thence at laſt, the *Navel-Fibres* ſhoot (as the privitive *Artery* into the

the *Colliquamentum*) through the *Secondine* into the aforesaid Liquor deposited therein.

Into which Liquor, being now shot, and its own proper Sap or Tinctures mixed therewith, it strikes it thus into a *Coagulum*; or, of a Liquor, it becomes a Body consistent and truly *Parenchymous*; and the supply of the said Liquor still continu'd, and the shooting of the Navel-Fibres, as is above described, still carried on, and the therewith said *Coagulation* or *Fixation* likewise.

And in the Interim of the *Coagulation*, a gentle *Fermentation* being also made, the said *Parenchyma* or *Coagulum* becometh such, not of any Constitution indifferently, but is thus raised (as we see Bread in Baking) into

into a *Congeries* of *Fixed Bubbles*: For such is the *Parenchyma* of the whole Seed.

FINIS.

THE
EXPLICATION
OF THE
FIGURES.

Fig. I.

Sheweth a Bean *with the two* Lobes *laid open somewhat wider than the Parts, without a Rupture, will well bear, for the better sight of that Part which lieth between them.*

aaaa The two Lobes.
AA Their contiguous Flats.
b The Radicle.
c The Plume.
dd One of the Cavities wherein the Plume lieth.

Fig.

Fig. 2.

aaaa The *Parenchyma.*

eeee The *ſeminal Root* diſtributed throughout the *Parenchyma* of either *Lobe.*

b The *Radicle*, with the *ſeminal Root* running through it in one Trunk to the Point thereof.

c The *Plume*, with the Diſtributions of its *Inner Body* continued from the *ſeminal Root* of either *Lobe.*

xx The oblique *Inſertion* of the two grand Branches of the *Lobes* into the Trunk of the *Radicle.*

Fig.

Fig. 3.

The Lobe *of a* Bean *cut athwart.*

aaa The convex or external part thereof.

bbb The concave ſide out of ſight.

cccc The Extremities of the Branches of the *ſeminal Root*, as they appear like ſo many ſmall Specks in the traverſe Cut.

Fig. 4.

The Plume cut athwart.

The black Specks represent the Branches of the *seminal Body* thereinto inserted, or therein distributed.

Fig. 5.

aaaa A *Lobe* of a *Gourd-seed.*
cccc The greater Branches.
ee The Sub-divisions and Inosculations of the lesser.

Fig.

Fig 5. OO.

AA A great white *Lupine.*

aa The *Navel-Fibres* which ſtrike from the Ramulets of the *ſeed-Branch*, into the *Lobes.*

ab The production of the *Navel-Fibre* into the *Radicle* (*b.*)

c The *Plume.*

bc The *Pith.*

aeeee The diſtribution of the *Navel-Fibre* in the *Lobes*; all becoming the *ſeminal Root*, deſcrib'd in the firſt Chapter.

Fig. 6.

aaaa A Slice of the Root of a Tree.

cccc The *Cortical Body* or *Barque.*

e The *Pith.*

The black Pieces are the Shootings of the *Lignous Body.*

The Specks therein are its *Pores.*

The White Pieces are the *Insertions* of the *Cortical Body.*

Fig.

Fig. 7.

Sheweth the Root of Berbery *in the Traverse Cut.*

aaa The *Cortical Body* or *Barque*. The white Lines are the *Insertions*. The Black Specks are the Pores of the *Lignous Body*.

Fig. 8.

aaaa The *Cortical Body* as appearing in a *Turnep* cut athwart.

acdacd The *Lignous Body*, or the ſeveral Shoots thereof repreſented in their Ranks, by the black Lines; the Pricks made along the Lines being the Terminations of the ſaid Shoots or Fibres; not viſible except in a thin ſlice, or after the Surface of the *Turnep*, being cut, is well dried.

cccc The *Cortical Body* inſerted betwixt the Shootings of the *Lignous*: or the *Pith*.

ab ab A piece of the *Cortical Body* taken off, that its own Inſertions (*eeee*) and the Oſculations of the *Lignous* may be ſeen; which is beſt done after the Inſertions are a little dried and ſhrunk.

Fig.

The Appearance of divers Roots, in their Elder estate, as ex. gr. *of a* Columbine.

Fig. 9.

he Fibrous parts of the *Root*, where the *Lignous Body* ſtands Central; the Pores whereof are repreſented by the black Specks.

10. The *Root* cut a little higher, where the *Cortical Body* ſometimes appears only once inſerted.

11. The *Root* cut higher with the Inſertions in ſome number.

12. The Inſertions ſtill more numerous.

13. The *Pith* (*a*) now begun, the ſaid Inſertions being collected in the Center.

14. The *Pith* (*a*) more amplified.

Fig. 15.

Sheweth a ſmall piece of the Trunk of Burdock.

a The juſt ſize thereof to the naked Eye.

aaaa The appearance of it through a *Microſcope.*

lll The Inſerted *Cortical Body.*

ccc The outmoſt ſhooting of the *Lignous Body* diſtributed into the Leaves.

ee bb tt The inner Shootings or Fibres diſtributed to the Branches.

The Black Specks are their Pores, which, through a *Microſcope* are fairly viſible in them all.

Fig.

Fig. 16.

aaaa The Slice of a Trunk of divers years growth.

cccc The *The Cortical Body*, or *Barque*.

e The *Pith*.

The white Lines are the Inſertions of the *Cortical Body* or *Barque*.

The Black Lines are the *lignous Body*.

The ſeveral Shootings thereof betwixt the black Circles ſhew the Annuall Rings.

Fig.

Fig. 17.

Sheweth a ſmall piece of Oak cut athwart.

b The juſt bigneſs of it, as it appeareth to the naked eye.

bbbb The appearance thereof through a *Microſcope.*

aaaa The greater Inſertions viſible to the bare eye.

The white Lines are the ſmaller Inſertions only viſible by the *Microſcope.*

cccccc The greater Pores viſible to the bare eye.

eeeeee The middle ſized.

The black Spots are the ſmalleſt of all, and both theſe latter viſible only through the *Microſcope.*

c The *Pith* of every great Pore.

Fig.

Fig. 18.

aaaa A piece of the Leaf of a Table.

bbbb The *lignous Body* with its Pores running by the length of the Trunk.

cccc The Inſertions of the *Cortical Body*, with the Tract of their Pores running directly croſs to thoſe of the *lignous*, *viz.* by the Diameter or breadth of the Trunk.

Fig.

Fig. 19.

A Slice of a younger Trunk of a Burdock.

cccc The utmost Shootings of the *lignous Body* contiguous to the Skin; wholly distributed into the outer Leaves.

eeee The middle Shootings running chiefly into the lower *Germens*.

et et &c. The inner Shootings belonging to the higher *Germens*.

a The *Pith*.

Fig.

The various Diſpoſure, Size and Figure of the Fibres *in the Stalk of a Leaf.*

Fig.

20 In *Endive* thus
21 *Coltsfoot.*
22 *Cycory.*
23 *Ivy.*
24 *Aſarabacca.*
25 *Mint.*
26 *Dock.*
27 *Borage.*
28 *Mullen.*
29 *Cabbage.*

FINIS.

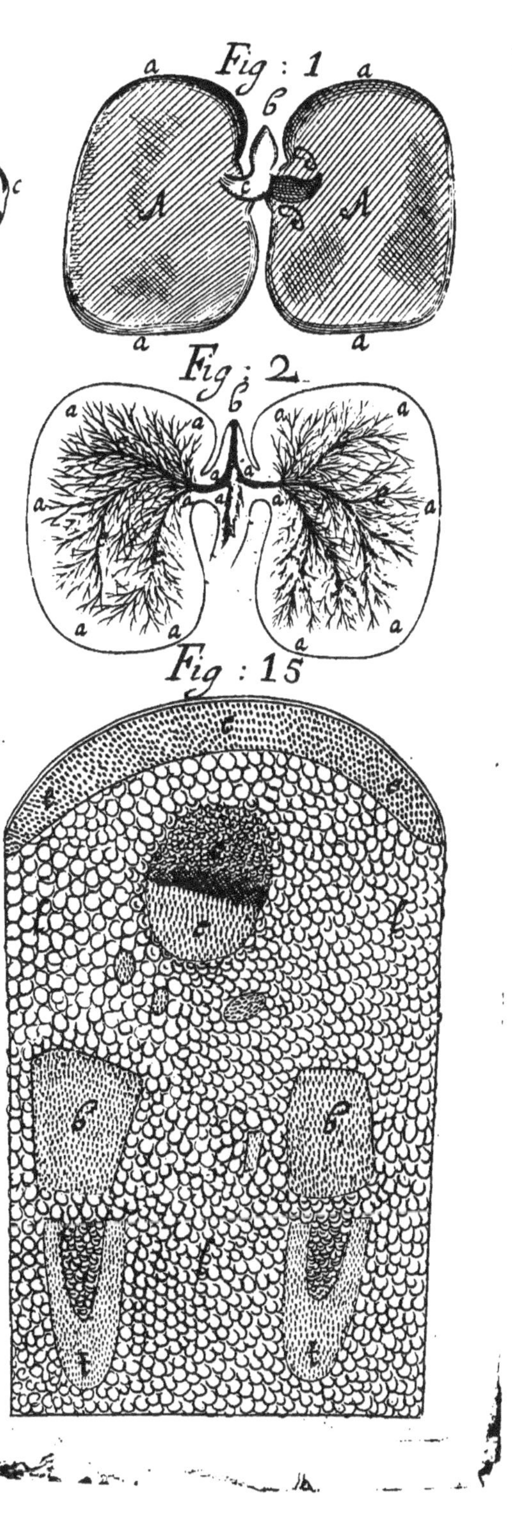
Fig : 1
a
a
b
A
A
a
a
Fig : 2
b
a
a
a
a
a
a
a
a
a
a
Fig : 15
c
c
c
b
b
t
t

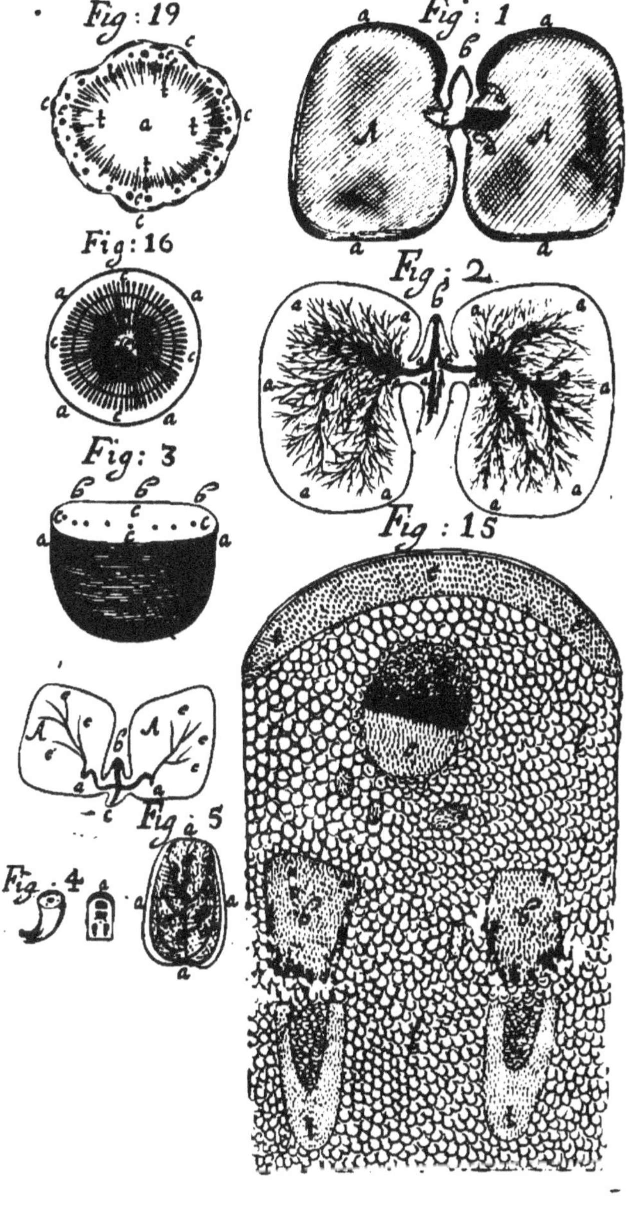
Fig: 19
Fig: 1
Fig: 16
Fig: 2
Fig: 3
Fig: 15
Fig: 5
Fig. 4

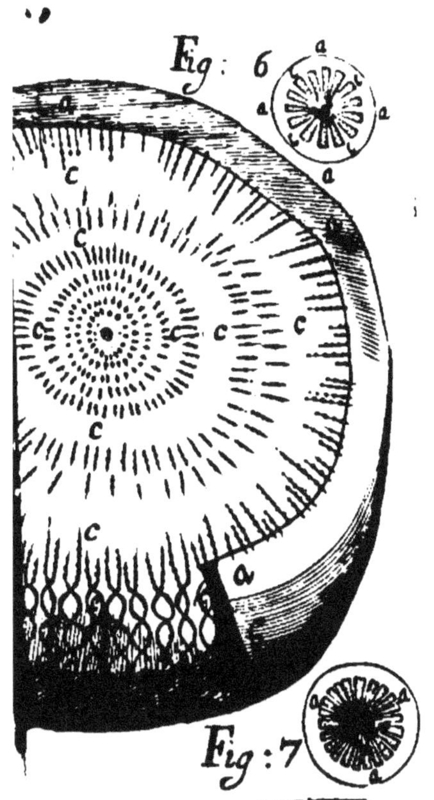

Fig: 6
a
a
a
a
c
c
c
c
c
c
c
a
Fig: 7
a
a
a

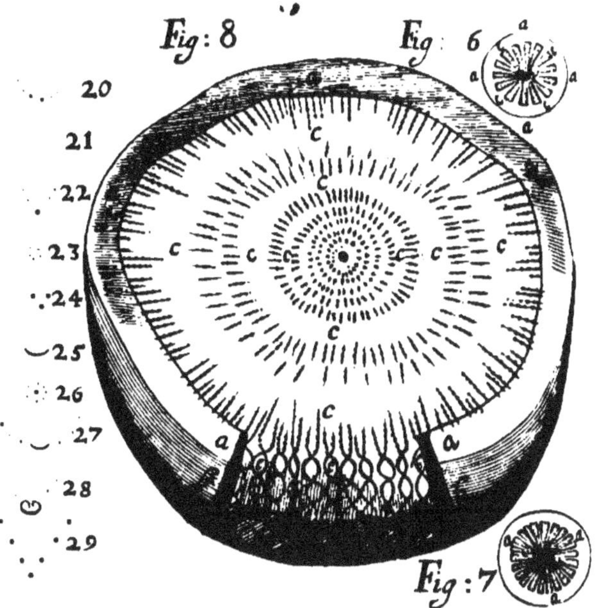
Fig: 8
Fig: 6
Fig: 7
20
21
22
23
24
25
26
27
28
29
a
c

www.ingramcontent.com/pod-product-compliance
Lightning Source LLC
Chambersburg PA
CBHW021526210326
41599CB00012B/1396

9783337163297